The resource for the
NEW AQA SPECIFICATION

Statistics
GCSE for AQA

Jayne Kranat
Part time teacher of Mathematics
and Statistics in Bromley, Kent.
Currently a senior examiner,
moderator, reviser and assessor
for AQA.

Brian Housden
Teacher advisor for the London
Borough of Barking and
Dagenham.
Currently Chief Examiner in
GCSE Statistics for AQA.

James Nicholson
Head of Mathematics at Belfast
Royal Academy.
Part time Research Fellow at
Queen's University of Belfast
developing statistics materials
for schools.

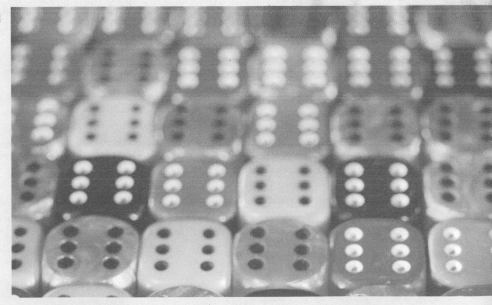

OXFORD
UNIVERSITY PRESS

D0319321

OXFORD

UNIVERSITY PRESS

Great Clarendon Street, Oxford OX2 6DP

Oxford University Press is a department of the University of Oxford.
It furthers the University's objective of excellence in research,
scholarship, and education by publishing worldwide in

Auckland   Bangkok   Buenos Aires   Cape Town   Chennai
Dar es Salaam   Delhi   Hong Kong   Istanbul   Karachi   Kolkata
Kuala Lumpur   Madrid   Melbourne   Mexico City   Mumbai
Nairobi   São Paulo   Shanghai   Taipei   Tokyo   Toronto

Oxford is a registered trade mark of Oxford University Press
in the UK and in certain other countries

The authors would like to thank
• David Hodgson for his invaluable contribution to this book.
• RoseMarie Gallagher and Anna King for use of their material on choropleth maps.

Database right Oxford University Press (maker)

First published 2001

British Library Cataloguing in Publication Data

Data available

ISBN 0 19 914788 4

The publishers would like to thank AQA for their kind permission to reproduce the Coursework Guidance and
Coursework Ideas page and the past paper questions. AQA accept no responsibility for the answers to the past paper
questions which are the sole responsibility of the publishers.

The publishers and authors are grateful to the following:

The Times, The Guardian, New Internationalist, The Express, The Daily Mail, Thomson financial datastream,
The Census at school project. Hulton Getty/Corbis for the use of the photograph on page 1.

Figurative artwork by Angela Lumley
Cover artwork by Photodiscs

Typeset in Great Britain by Tech-Set Ltd.
Printed and bound in Italy by Canale.

About this book

This book has been written by experienced examiners specifically for the AQA GCSE Statistics Specification 3311.

The book is organised into nine units containing all the up to date content and methods you will need in your exam.

How to use this book

The first seven units give you the content of the specification and broadly follow the same order as the specification.

Each of the content sections starts with a list of learning objectives. This will help you plan your work effectively. The **Before you start** section gives a list of prior knowledge you need before you start the unit and **Check in** questions to check you know the methods.

Shaded boxes in the text highlight important information you need to remember:

> Qualitative data describes a quality that cannot be counted.

Each unit ends with a **Summary** page detailing what you should know after studying the unit and providing **Check out** questions which are useful for revision at the end of the course.

At the end of the section there is a **Revision exercise** containing past paper questions from AQA Statistics papers. They will help you get ready for the style of question you will see in the actual exam.

At the end of the book there are **Practice exam papers** written by experienced examiners to help you prepare for the real thing.

Unit 8 is about **Coursework**. Work through this unit carefully before you start your coursework and it will help you make the most of this important element.

Unit 9 is about using **ICT in Statistics**. It will show you how to use your calculator more effectively and how to make the most of the Internet as a research tool. Finally it will encourage you to use Spreadsheets, particularly for your Coursework.

To help you make the most of the ICT opportunities in this unit there are **icons** throughout the book referring to the relevant pages in Unit 9.

Colour coding

Blue and grey are used to show whether the material is Foundation or Higher or both.

Foundation	Higher	Both

You will find the coding in:
- ✦ Section headings
- ✦ Key points
- ✦ Examples
- ✦ Exercises
- ✦ Questions

Remember that all the Foundation content is also examined in the Higher tier.

GCSE Statistics for AQA Contents

▪ This symbol shows you the content of the GCSE Statistics specification that is not in the GCSE Mathematics specifications.

1 Data collection

We are just statistics, born to consume resources.
Horace (65–8 BC) Epistles Book 1

This unit will show you how to

- ✦ Classify data
- ✦ Take a sample of data
- ✦ Design questionnaires and carry out a survey
- ✦ Conduct experiments
- ✦ Draw tables to collect data

Before you start

You need to know how to	Check in 1

1 Use tally marks
| is a tally mark.
|||||||||||||||| would be difficult to count.
Counting in fives is easier, so group tally marks as a five-bar gate ⅢⅠ.
||||||||||||||||| is ⅢⅠ ⅢⅠ ⅢⅠ || = 5 + 5 + 5 + 2 = 17.
	1 (a) How much is
	   (i) ⅢⅠ ⅢⅠ ⅢⅠ ⅢⅠ ||| (ii) ||||
	   (iii) ⅢⅠ ⅢⅠ?
	(b) Write using tally marks
	   (i) 6, (ii) 14, (iii) 20, (iv) 2, (v) 5

2 Round numbers
  ✦ Round down numbers ending in 0, 1, 2, 3 or 4
  ✦ Round up numbers ending in 5, 6, 7, 8, or 9.
8.4 cm rounds to 8 cm.
8.6 cm rounds to 9 cm.
	2 Round to the nearest whole number
	(a) 4.1, (b) 18.9, (c) 23.0,
	(d) 10.5, (e) 33.3, (f) 66.6

3 Use proportion A proportion can be like a fraction. In a bag of 20 sweets, 7 are orange. The proportion of orange sweets is $\frac{7}{20}$.	**3** A bag of 40 marbles has 10 green, 7 purple, 8 yellow, 3 white, 6 blue and 6 red. What proportion of the marbles are (a) red, (b) green, (c) white, (d) yellow, (e) yellow or white, (f) blue or green, (g) not purple?

1.1   Types of data

Vast amounts of raw data are being collected all the time.

Raw data is information that has not been ordered or processed in any way.

Data can be

qualitative (non-numerical)
For example: the texture and colour of a fabric are properties that are not numbers.

quantitative (numerical)
For example: the number of books in a room or the height of a person are numbers.

Qualitative data describes a quality that cannot be counted.

Quantitative data describes a quantity that can be counted or measured.

Quantitative data can be

discrete (countable)
For example: the number of rooms in a house, or a person's shoe size.

continuous (measurable)
For example: length, weight and time are all measured on a continuous scale.

Discrete data is usually concerned with a limited number of countable values.

Continuous data is measured on some scale and can take any value within that scale.

> Continuous data may sometimes seem discrete. For example, you may record time to the nearest second.

You usually collect data on a particular **variable.**

A variable is a property able to assume different values.

For example, temperature is a variable and data can be collected on it.

Example

Imogen buys a new dress.
Write down two variables associated with a dress that illustrate the
following data types: (a) qualitative, (b) discrete and (c) continuous.

(a) Colour and texture of the material are qualitative.

(b) The size of the dress and the number of buttons it has are
    discrete.

(c) The length of the sleeves and the diameter of each button are
    continuous.

Exercise 1A

1. Robert is spending the summer hiring out deckchairs at the beach.
   (a) Is the number of deckchairs hired out each day a discrete or
       continuous variable?
   (b) Describe a qualitative variable associated with the deckchairs.

2. Sheena is buying a wig for a fancy dress party.
   Describe a qualitative variable associated with the wig.

3. Adam takes 23 minutes to complete a jigsaw puzzle.
   Is this a discrete or continuous variable?

4. Give an example of (i) a qualitative and (ii) a quantitative measure
   associated with each of the following.
   (a) A toy car.
   (b) A shoal of fish.
   (c) A computer game.

5. Sort the following into (i) discrete and (ii) continuous data.
   (a) How long you take to finish in a cross-country race and your
       finishing position in the race.
   (b) The weight of a parcel and the cost of its postage.
   (c) The number of eggs and the amount of sugar needed in a cake
       recipe.

6. David attended a job interview. His communication skills were rated
   as either 1 – good, 2 – average or 3 – poor.
   Why is this qualitative and not quantitative data?

7. Gareth is looking at the night sky through a telescope. Is the
   number of stars in the galaxy discrete or continuous data?

1.2   Collecting data

Before you collect data you should have an aim in mind.

You then need to decide what sort of data to collect and the most appropriate and efficient method to obtain it.

Secondary & primary data

You can use data that someone else has collected and published or you can collect it yourself.

> Primary data is data that you collect yourself.

> Secondary data is obtained from published statistics. It is data that already exists, so it is second hand.

> Secondary data sources include: Social Trends, Economic Trends, Annual Abstract of Statistics, Monthly Digest of Statistics, newspapers and the Internet.

The table shows the man advantages and disadvantages of using primary and secondary data.

Data	Advantages	Disadvantages
Secondary	Cheap Easy to obtain May give a starting/reference point for a survey or experiment.	Could be unaware how it was collected, who it was collected from, how it was collated. May be out of date.
Primary	You know how it was collected. You know who it was collected from.	Expensive Time-consuming

Experiments & surveys

> You can use an experiment to collect data.

Experiments are particularly useful for collecting scientific data.

In an experiment, at least one of the variables is controlled – this is called the **explanatory** or **independent variable**. The effect is observed – this is called the **response** or **dependent variable**.

Example

Edith carries out an experiment to find out which age group at a school is best at estimating a minute.

Suggest independent and dependent variables for her experiment.

She chooses the age groups so the age is the independent variable. She records the responses, so the time estimated is the dependent variable.

> You can use a survey to collect data.

Surveys are particularly useful for collecting personal data.

The main survey methods are:

+ Observation, which involves monitoring behaviour or information.

+ Personal interviews, which are widely used in market research.

+ Telephone surveys, which are a special type of personal interview.

+ Postal surveys in which the survey is sent to an address.

The table outlines the main advantages and disadvantages of these survey methods.

Survey method	Advantages	Disadvantages
Observation	Systematic & mechanical.	Results are prone to chance.
Personal Interview & Telephone Survey	High response rate Many questions can be asked.	Expensive The interviewer may influence the replies.
Postal Survey	Relatively cheap Large amount of data can be collected.	Poor response rate Limited in the type of data that can be collected Only part of the survey may be completed.

Example

A car design company wishes to design and market a new sports car.

(a) Explain how and why they could use both secondary and primary data.

(b) Which method of collecting primary data could be used?

(a) Secondary data – you could refer to published data on human body measurements, to decide on roof height, seat/gear stick distance etc.
Primary data – conduct a survey to find out which features in a car would be most attractive to your market.

(b) The survey could be conducted by personal interview or by post.

Exercise 1B

1. To improve the road safety in Dangercity the council propose to install speed restriction humps. They study the accident rates for the busiest roads.
Are they using primary or secondary data?

2. Darryl and Jason want to predict next season's football league champions. Darryl looks at the football results from 2000. Jason looks at the results from 1995 to 1999.

   (a) What type of data are they using?
   (b) Whose opinion is likely to be more reliable and why?

3. Choc-u-like is a confectionery company. They want to produce a new chocolate bar.
   Should they collect primary data or use secondary data?
   Give a reason for your answer.

4. Coftea is a manufacturer of kettles.
   How could they use secondary data in designing a new kettle?

5. Jasmine is convinced that people with longer legs run faster in sprint races. She conducts an experiment to test her theory.
   What are the explanatory and response variables she should measure?

6. Richie Vanson, a businessman, is considering building a leisure centre in Tintown.
   What method of collecting primary data should he use to help him decide whether to build?
   Give a reason for your answer.

7. A survey is carried out to investigate the cost of heating newly built houses.
   Give one reason why this survey should be carried out by post.

1.3 Sampling

It depends on who is included in a survey as to the picture the survey gives.

When you want to collect data you must first identify your **population**, that is, who or what you want to include in your survey.

> A **population** is everything or everybody involved in the study.

Census data

> Census data obtains information from every element of a population.

An example of a census involving a large population is the Government National Census conducted every ten years.

✦ A census is usually only practical when the population is small and known, for example when studying the working conditions of a small firm or the television preferences of a class of pupils.

✦ A census is not always possible, for example if you wanted to find the weight of all the tea in China.

Census information is used by central and local government, health authorities and many other organisations to allocate resources and plan services for everyone.

Sample data

When a census is not practical or possible you need to take a sample of the population.

The purpose of sampling is to collect data from some of the population and use it to make conclusions about the whole population.

Two samples drawn from the same population are likely to give different results.

> To collect sample data, you take information from part of the population.

You should try to ensure that the sample you choose is free from **bias**.

Bias in sampling may arise from:

✦ Misidentifying the population.
✦ Choosing an unrepresentative sample.
✦ Non-response to a survey.
✦ Asking ambiguous or leading questions.
✦ Dishonesty of those sampled.
✦ Errors in recording answers.
✦ Not controlling external factors in experiments, such as the effect of noise on the ability to complete a task requiring concentration.

In statistics, bias means a distortion of results.

An external factor is also called an extraneous variable.

The main advantages and disadvantages of using census data or taking a sample from a population are outlined in the table.

	Advantages	Disadvantages
Census	Accurate Unbiased Includes every item in the population	Expensive Time-consuming Difficult to ensure the whole population is surveyed
Sample	Cheaper Less time-consuming Reduces the amount of data to be collected and analysed	May be biased Not totally representative

Exercise 1C ——————————————————————————

1. Citsale, elastic band producers, conducted an experiment to see how much a new elastic band would stretch.
   Identify the population to use in the experiment.
   Give a reason why a census should not be carried out.

2. A new canteen was being built at a small factory.
   Give a reason for the management to conduct a census survey to determine the factory canteen's menu.
   Identify the population for this survey.

3. For the following studies, decide whether you would conduct a census or a sample survey. Give a reason for your choice.
   (a) A study of house prices in England and Wales.
   (b) A study of house prices in the village of Smallbury, population 105.

4. A school proposes to install floodlights on their football pitch.
   Identify the population affected by their proposal.

1.4   Sampling methods

To ensure each item in the population has an equal chance of being selected, you take a **random sample** from the whole population.

> A random sample is one chosen without conscious decision.

Usually, the larger the sample the more representative the statistics are of the whole population.

A **sampling frame** is used to identify the population. It consists of all items in the population and ensures that they all have a chance of being sampled.

Often there has to be a balance between practical convenience and ideal conditions. To obtain a true random sample you use simple random sampling.

Simple random sampling

Every item in the population has an equal chance of being chosen. You assign a number to each item and then use random number tables, a calculator or computer to choose a sample, ignoring duplicate numbers.

This method is more suited to a relatively small population where the sampling frame is complete.

Although using random numbers is free from (personal) bias, there is no guarantee that the actual sample chosen is unbiased.

Example

Below is an extract from a table of random numbers.

Use the numbers to identify a sample of 5 from a population of 50.

67485 88715 45293 59454 76218 78176
87146 99734 35555 76229 00486 64236
74782 91613 53259 79692 47618 20025
16022 27081 00058 58042 67833 23539
37668 16324 97243 03199 45435

- ✦ Assign a number to each member of the population from 00 to 49.
- ✦ Choose any point in the table to start and read pairs of numbers in a direction of your choice.
- ✦ Ignore repeated numbers and numbers above 49.

- ✦ For example, start here and read across.
- ✦ Then the pairs you read are:
  ~~70~~ ~~84~~ 00 05 ~~85~~ ~~80~~
  42 ~~67~~ ~~83~~ 32 35
- ✦ The crossed out numbers are too big for the question, so your sample of 5 would be:
  0, 5, 42, 32 and 35.

Stratified sampling

The population is divided into categories (strata) by age, gender, social class..., then a random sample is chosen from each category. The size of each sample is in proportion to the size of each category within the population.

> Stratified sampling ensures you have a fair proportion of responses from each group of the population.

For example in an infant school with 150 pupils where 30 are in reception year, 60 in year 1 and 60 in year 2; a stratified sample of size 15 would choose 3 from reception year, 6 from year 1 and 6 from year 2.

Systematic sampling

A regular pattern is used to choose the sample. Every item in the population is listed, a starting point is randomly chosen and then every nth item is selected.

For example every fiftieth packet of seeds from a production batch is tested for germination starting with the eighth.

This is a simpler and quicker method to select a (random) sample, but may be unrepresentative if a pattern exists in the list.

I'll choose 30 people from each year – 15 boys and 15 girls.

Cluster sampling

The population is divided into groups or clusters. A random sample of clusters is chosen and every item in the chosen cluster is surveyed.

A large number of small clusters minimizes the chances of this being unrepresentative.

This method is used for example by biologists to study flora and fauna.

Quota sampling

Instructions are given concerning the amount (quota) of each section of the population to be sampled.

A disadvantage is that the actual people or items chosen are left to the discretion of the surveyor which could lead to bias.

An advantage of this method is that no sampling frame is required.

This method is commonly used in market research.

> If there are n items in the population, an appropriate sample size is $\sqrt{n}$.

> To quarter $\left(\frac{1}{4}\right)$ the variability you use $4^2 = 16$ times the sample size.

Convenience sampling

The most convenient sample is chosen which, for a sample of size sixty, could mean the first sixty people you meet.

It is highly likely that this sample would be biased and unrepresentative.

Opinion polls

Large-scale opinion polls often use a combination of cluster and quota sampling.

An example of this is the accurate estimates of the outcome of general election for public office.

The sample size may be **large** but is often based on a very small proportion of the population.

A major disadvantage of conclusions drawn from opinion polls is that opinions may change over time.

> The criteria for selection of a sample in national opinion polls are: geographical area, age, gender, social and economic background.

Example

Nyree wanted to investigate whether people measured their height using metric or imperial units. She went to her local supermarket and asked the first twenty people she saw how tall they were.

(a) For this survey state the sampling frame, the sampling method used and why it might be biased.

(b) Outline a better method to use to choose the sample.

(a) The sampling frame could be the whole population of this country (or the whole population in the world).

The sampling method used is convenience sampling. It is biased as everyone chosen most probably lives in the same area and so may not provide a cross-section of social class. Also the sample is very small.

(b) A better method could be to use quota sampling, ensuring for example that men and women across all ages, children, different ethnic groups and different social classes were all represented.

> If Nyree had wanted to know how people in her town measured their height, then she could have used stratified sampling as the sampling frame would be every person living in that town.

Exercise 1D ⎯⎯⎯⎯⎯⎯⎯⎯⎯⎯⎯⎯⎯⎯⎯⎯⎯⎯

For each of the following questions:

(a) identify the population (sampling frame),
(b) explain why the sample may be biased,
(c) explain a better method to use to choose the sample.

1. John was carrying out a survey to find how far, on average, residents in his town travel to work. He asked all the people at his local railway station one Monday morning.

2. Hazel thinks that boys at her school get more pocket money than girls. There are 300 children at the school, 120 boys and 180 girls. In her survey she asks 30 boys and 30 girls.

3. To find out attitudes on abortion, an interviewer stopped people in a local shopping centre one weekday morning and asked shoppers their views.

4. Pedro wanted to find out how much people in Britain were prepared to spend on holidays abroad. He asked people on the street where he lives.

5. Catriona believes that more people in Scotland get married in church than in a registry office. She asks all the people attending a church service where they got married.

6. To investigate the statement 'children no longer do enough sport', all the children at one school in Downtown were surveyed.

7. Glowalot, a light bulb manufacturer claimed that their light bulbs lasted for more than 200 hours. Gina thought it would be a good idea to test their claim by lighting all the bulbs produced in one month.

8. Joanne wanted to find out if rich people in England smoked more cigarettes than poorer people. She chose 200 households at random and sent them a 40 page questionnaire which they had to pay postage to return.

9. Larry decides to estimate the number of blades of grass in his lawn. He stands on the lawn and counts the blades of grass within 40 cm of his feet.

10. A machine producing split pins is believed to produce defective pins at a rate of 10%. A systematic sample was chosen to test this.

1.5 Questionnaires

You often need to use a **questionnaire** to collect data.

> A questionnaire is a set of questions used to obtain data from individuals.

The questions can ask for factual information, for example 'how many radios do you own?' or for opinions, for example 'which radio station do you prefer to listen to?'

People who answer questionnaires are called **respondents**.

The questionnaire itself should:

+ Be clear as to who should complete it.
+ Start with easy questions as this encourages the respondent to continue.
+ Be clear where and how the answers should be recorded.
+ Be able to be answered quickly.
+ Be as brief as possible.

Individual questions should:

+ Be short.
+ Use simple and clear language.
+ Be unambiguous.
+ Be free from bias.
+ Not be leading (questions of the type 'do you agree that...' lead the respondent to agree).
+ Be useful and relevant to your survey.

There are different questioning techniques you can use:

Open questions

> An open question has no suggested answers and allows a respondent to reply with a single word or a long explanation.

Open questions may reveal responses or opinions that you have not considered.

Closed questions

> A closed question has a set of suggested answers to choose from.

The advantage of using closed questions is that it is easier to summarize all the data obtained and make comparisons.

Tick box choices for closed questions

The choices offered in a closed question should not overlap or leave gaps.

For example if you wanted to find out the age group of your respondent and the choices were:

☐ under 10   ☐ 11–20   ☐ 21–30   ☐ 30–40   ☐ over 40

✦ a 10-year-old would not be able to answer.
✦ a 30-year-old will have two answer boxes.

Opinion scales

If you use an opinion scale, responses tend to cluster around the middle of the scale as people do not want to appear extreme.

For example if the answer choices were given as tick boxes:

☐   ☐   ☐   ☐   ☐
strongly disagree   disagree   no opinion   agree   strongly agree

or respondents were asked to mark on a scale:

├────────────────────────────────────┤
disagree   agree

the majority of respondents would choose the middle options.

> One way to avoid this is to provide an even number of options so that there is no middle choice.

Pilot surveys

> **A pilot survey is a small replica of the actual survey (or experiment) that is to be carried out.**

The pilot should help you identify potential problems with the wording of the questions, and so limit the errors in expensive full-scale surveys.

If you ask closed questions in a survey, you must ensure that the choices of answers reflect the typical answer of the respondents.

To do this, use open questions in a pilot survey to find out the kinds of responses you will get. Then use closed questions in the actual survey.

> A pilot survey may also give an estimate of the non-response rate and a guide as to the adequacy of the sampling frame.

Sensitive questions

Few people willingly provide intimate facts about themselves and many resent being asked such questions.

If you need to ask highly personal questions, you should leave them to the end of a questionnaire.

To ensure truthful answers to sensitive questions you can use the 'random response' technique:

Example

'Toss a coin. If it shows heads tick the yes box; if it shows tails answer the question:

Have you ever used class A drugs? ☐ yes ☐ no'

If 20 people answer the question then the expected number of 'heads' is 10. If 13 of the people answer 'yes' the proportion who 'genuinely' answer 'yes' is $\frac{3}{10}$.

There is more about expected outcomes on page 226.

Exercise 1E

Comment critically on questions 1 to 6.
Suggest how they could be improved, either by rephrasing the question or by giving a choice of answers.

1. Do you watch a lot of television?

2. Do you agree that the teachers at your school are superb?

3. Approximately how tall are you?
    ☐ under 1 metre   ☐ over 1 metre, but less than 5 feet
    ☐ 2 metres or more

4. (a) What do you think of the facilities at the new leisure centre?
    (b) Have you visited the new leisure centre? ☐ yes ☐ no

5. How old are you? ☐ young ☐ middle-aged ☐ old

6. How many times during the last week did you take a bath or shower?
    ☐ once ☐ twice ☐ every day

7. Write a questionnaire to find out the most popular flavour of crisps for different age groups in your school.
    State (a) the sampling frame
          (b) the sample size and how you would obtain the sample.
    Describe how you could use a pilot survey.

8. A manufacturer of CD players has asked you to find out the features people would want on a portable personal CD player and how much they would be prepared to pay.
    Write a questionnaire for a pilot enquiry to find this information.
    State (a) the sampling frame for the whole survey
          (b) how you would choose the sample for the pilot enquiry.

1.6 Experimental design

If you carry out an experiment as part of your investigation, here are some methods that you should consider:

Data logging is a mechanical or electronic method of collecting primary data. You programme a machine to take readings at set intervals.

You use data logging to measure rates such as your pulse rate after exercise.

You use a **control group** when you want to test the effect of different factors in an experiment. A randomly chosen group, the control group, is not subjected to any of the factors that you wish to test.

To test how nutrients affect the growth of plants you should have a control group that is not given any nutrients.

Other experimental methods

You can use **matched pairs**, using two groups, to investigate the effect of a particular factor. Both groups to be tested need to have everything in common except for the factor to be studied.

In the 1930s, Newman investigated the effect of the environment on personality through studying two identical twins reared apart, from an early age, in different environments.

You could conduct a **before-and-after experiment** to help judge the influence that a factor could have.

Schools determine the effect of the long summer break on childrens' education by testing at the end of one school year and then again at the start of the following year.

You can use a **capture–recapture method** to estimate the size of a self-contained population. A sample is taken and tagged and then returned to the population. Some time later a second sample is taken and the number of tagged members is noted. An estimate of the total population can then be made.

For the estimate to be reliable, there needs to be extensive intermingling of the first sample within the whole population.

Example: capture–recapture method

Forty ducks are caught, tagged and returned to a bird sanctuary. In a second sample of sixty ducks, twelve are found to have tags from the first sample.

Estimate the number of ducks in the sanctuary.

The proportion $\frac{12}{60}$ is estimated to be the same as $\frac{40}{n}$ where n is the number of ducks in the sanctuary.

$$\frac{12}{60} = \frac{40}{n}$$

$$n = \frac{40 \times 60}{12}$$

$$n = 200$$

Exercise 1F

For each of the experiments numbered 1 to 7, decide which experimental method could be used.

1. A new treatment for malaria is to be tested for effectiveness.

2. The heart rate of a sample of people of differing ages is to be monitored as they perform a particular physical exercise.

3. A dormant volcano is monitored for activity.

4. The growth of a genetically modified crop is monitored.

5. An estimate is to be made of the number of deer in a particular area of woodland.

6. An assessment is to be made on the effects of alcohol on driving ability.

7. The population of Jackass penguins, an endangered species, is monitored for change in population.

8. Fifty fish are caught, tagged and returned to a particular pond. In a second sample of thirty fish, five are found to have tags.
   Use the capture–recapture method to estimate the number of fish in the pond.

9. There was a population explosion at a rabbit sanctuary. 140 rabbits were caught, tagged and returned to the sanctuary. A second sample of 80 rabbits were snared. 28 of these were tagged. Estimate the number of rabbits in the sanctuary.

1.7   Presentation of data

You collect raw data by counting or by measuring. It is useful to collect the data in a frequency table so that it is easier to analyse.

> You can use tally marks to record data in a frequency table.

Ahmet wanted to find out the pets owned by his classmates.
Here is part of his frequency table:

Pet	Tally	Frequency
Dog	IIII	4
Cat	JHT	5
Goldfish	JHT III	8
. . . . . .		. . .

Add up the tallies to find the frequency

List the categories on the left, leaving space for unusual responses.

Collect the responses in 5s to make the data easier to count.

Exercise 1G

1. Draw frequency tables to find
    (a) the frequency of type of pet owned in your class
    (b) the colour of bicycle owned by your classmates.

2. A survey was carried out to find the number of gel pens owned by each member of a class, with the following results:

    9 8 12 15 9 13 14 13 13 11 12 10 9 9 14 8 11 10 9 12

    Draw a tally chart to summarize this data.

3. Throw a dice 30 times and record the outcomes in a tally chart.

4. Draw a tally chart to record the number of times each vowel –
    a, e, i, o and u – appears on page 18 of this book.

1.8   Limits of accuracy

When you collect continuous data it will be rounded to the nearest sensible measurement.

> The **upper bound** is the largest value that a measurement could be.
> The **lower bound** is the smallest value that a measurement could be.

Age is an exception to this rule. If you are 15 years and 8 months you are still aged 15 despite being closer to 16 years old.

Example

The width of my computer screen is 37 cm.

This measurement is given as a whole number (to the nearest integer).

The *actual* measurement of my computer screen must be in between 36.5 cm and 37.499 999 . . . cm.

36.5 cm is the lower bound and 37.499 999 . . . cm is the upper bound.

If you measure more accurately and find the width of the computer screen is 36.7 cm:

the lower bound is 36.65 cm and the upper bound is 36.749 99 . . . cm.

> In practice, you would say that the upper bound is 36.75 cm.

Exercise 1H

Find the upper and lower bounds for questions 1–6.

1. The cost of a CD player given as £120 to the nearest pound.

2. The cost of a CD player given as £120 to the nearest £10.

3. The weight of a box of cereal given as 750 g to the nearest gram.

4. The weight of a box of cereal given as 750 g to the nearest 10 g.

5. The weight of a bag of flour given as 1 kg to the nearest 20 g.

6. The distance between two towns given as 45 km to the nearest 5 km.

> Remember that the smallest unit of money is 1p.

What are possible upper and lower bounds for questions 7–10?

7. The waistband of a skirt labelled as 26 cm.

8. A garden path measured to be 40 m long.

9. A tree measured to be 9.7 m high.

10. A beetle weighed as 35 g.

1.9   Grouped frequency tables

You can use a grouped frequency table to record your data.

To record data on foot length, part of it may look like this:

Foot length in cm	Tally	Frequency
. . . . . .		
22		
23		
. . . . . .		

The foot length 22 cm is measured to the nearest cm and so actually represents all foot lengths from 21.5 cm to 22.5 cm.

In a grouped frequency table you collect data using class intervals (the lower and upper bounds).

Foot length in cm	Tally	Frequency
. . . . . .		
21.5 and up to 22.5		
22.5 and up to 23.5		
. . . . . .		

You can write the class intervals just using the lower or upper bounds, that is

Foot length in cm   or   **Foot length in cm**

. . . .

21.5 –   – 22.5

22.5 –   – 23.5

. . . .

> Once you enter a tally mark in a class, the raw data (the actual measured length) is lost.

When you design a grouped frequency table you need to decide on the size of each class interval.

Note that:

+ Each class interval does not have to be the same width.
+ There should be no gaps or overlaps between the class intervals.
+ You will always lose some detail when combining two or more categories.
+ A grouped table may be more difficult to read.
+ Discrete and continuous data can be collected in a grouped frequency table.

Collecting raw data in a grouped frequency table

When you collect raw data in a grouped frequency table you can decide on the width of the class intervals. However,

+ Over-simplification (data falling into too few groups) effectively loses trends.
+ Under-simplification loses detail without highlighting trends.

The first or last class interval in a table could be open-ended (that is it has no lower bound or upper bound respectively), but this may make some calculations impossible to perform and some diagrams difficult to draw.

The most common reason to leave a class open-ended is because to specify a limit may be misleading.

Exercise 1I

1. The lifetime in hours of 40 electric bulbs was recorded as follows:

   34 48 76 56 58 64 30 72 63 58 76 51 50 43 44 67 78 64 32 39
   61 78 65 43 48 59 42 64 57 51 78 30 79 67 64 39 44 53 55 42

   Draw a grouped frequency table using equal class widths 30–39,
   40–49, ..., 70–79.

2. The times taken, in seconds, for a group of children to swim
   25 metres were:

   25.4 31.1 19.7 26.0 29.9 32.2 21.7 24.1 28.6 27.1 19.2 27.2 25.9 23.8
   25.5 30.1 20.4 26.1 28.7 21.4 29.6 30.5 33.1 26.9 27.0 21.9 24.6 26.9

   Draw a grouped frequency table using equal class widths 19–21,
   22–24, ..., 31–33. (Think about the lower and upper bounds of the
   class intervals.)

3. Measure the height of each member of your class.
   Design a frequency table to summarize these data.

1.10 Bivariate data and two-way tables

p266
p269

In many surveys and experiments you look at links between variables
and so data is collected for two variables, for example height and
weight, age and reaction time.

This type of data is **bivariate** and can be summarized in a **two-way table**.

Example

The following table shows the distribution of numbers of brothers and
sisters from a survey of a group of children.

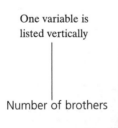

One variable is
listed vertically

One variable is listed horizontally

Number of sisters

Number of brothers

		0	1	2	3
	0	4	6	3	1
	1	7	5	3	0
	2	1	0	2	0
	3	1	0	0	0

3 people have 2 sisters
and 1 brother

(a) How many children took part in the survey?
(b) How many children have (i) no brothers or sisters?
                           (ii) one brother, but no sisters?
(c) How many children are there in the largest family?

(a) To find how many children took part in the survey, you need to find the sum of each of the entries in the table, that is

$4 + 6 + 3 + 1 + 7 + 5 + 3 + 1 + 2 + 1 = 33$

(b) (i) There are 4 children who have no brothers or sisters, you look for the number in row 0 and column 0.

   (ii) To find the number of children who have one brother, but no sisters you need to look in row 1 and column 0. The answer is 7 children.

(c) The largest family is the one where there are the most brothers and sisters. In two families there are 4 siblings (2 brothers and 2 sisters). There are 5 children in these families.

Exercise 1J

1. The principal of a music school posted letters to all of his students. The number of days it took for the letters to arrive is summarized in the following table:

Number of days / Postage type	1	2	3	4	5
First Class	12	4	2	0	1
Second Class	0	5	6	4	3

(a) How many students are there at the music school?
(b) How many of the first class letters took more than one day to arrive?
(c) How many letters took 3 days to arrive?

2. The weight and mint year of a sample of ten pence pieces are listed below.

1998–10.3 g	1996–9.8 g	1996–10.0 g	1998–10.5 g	1999–11.0 g
1999–10.8 g	1997–9.9 g	1997–10.2 g	1999–10.9 g	1996– 9.6 g
1997–10.5 g	1996–9.8 g	1996–20.2 g	1998–10.3 g	1998–10.4 g

The weights were measured to the nearest tenth of a gram.
Why do you think that the 1996 coin weighing 20.2 g may have been incorrectly recorded?
Ignoring this coin, copy and complete the table.

Weight in g	Year			
	1996	1997	1998	1999
9.6–10.0				
10.1–10.5				
10.6–11.0				

Comment on the results.

3. Carry out a survey to find out the number of brothers and the number of sisters each person in your class has.

4. Design a two-way table to find the types of CDs boys and girls in your class have bought in the last three months.

5. It is thought that the weight of a fruit pastille is linked to its colour. Design a two-way table to collect data to test this idea. (You may want to carry out this experiment.)

Summary

You should now be able to	Check out 1
1 Classify data.	1 Give an example of (i) a qualitative and (ii) a quantitative variable associated with a mobile phone.
2 Take a sample of data.	2 A school with 400 girls and 600 boys wants to find out the most popular features on a mobile phone. Describe how they could sample the opinions of their pupils.
3 Design questionnaires and carry out a survey.	3 Write out a question for this survey.
4 Conduct experiments.	4 The majority of pupils surveyed said their phone calls lasted for 30 seconds. Describe an experiment you could carry out to see if the pupils know how long 30 seconds is.
5 Draw tables to collect data.	5 Design a grouped frequency table to collect this data.

Revision Exercise 1

1. A doctor records information on her patients.
   The variables she uses are described below.
   State whether each variable is qualitative, discrete or continuous.
   (a) the colour of the patient's eyes
   (b) the patient's weight
   (c) the patient's shoe size
   (d) the patient's blood group   [NEAB]

2. Copy and complete the table by naming the type of data formed by each of the stated measurements.
   The first one has been completed for you.

Measurement	Type of data
Height of rose trees	Continuous
Number of brothers	
Length of shoe laces on pairs of shoes	
Number of pages in library books	

[SEG]

3. Mr and Mrs Jones were carrying out a survey.

They asked members of their sports club how much money they spent on sport, how many children they had and how long it took them to travel to the sports club.

(a) What kind of data source is this?

(b) Give an example of a **discrete** variable that Mr and Mrs Jones collected.

(c) Give an example of a **continuous** variable that Mr and Mrs Jones collected.   [SEG]

4.

The picture shows some of the ingredients needed for making a fruit cake.

Give **one** example of **each** of the following to do with making a cake:

(a) a continuous variable   (b) a discrete quantitative variable

(c) a qualitative variable   [SEG]

5. Lisa carried out a survey to find out people's views on music. These are the two questions that she asked:

1. Do you prefer pop music or classical music?
2. Are you aged under 25?

Design a two-way table to summarize her results.   [NEAB]

6. In a restaurant, customers have a choice for each of the three courses.

First course	Fruit juice or Melon
Second course	Fish or Chicken or Vegetarian
Third course	Sweet or Coffee

The waitress has to record the choices of 6 customers.

(a) Design a table for recording the choices if all that is needed is the total number of people wanting fruit juice, the total number wanting melon etc.

(b) Design a table for recording the choices of each of the customers.   [NEAB]

7. At a pilot training centre, twenty new pilots were given reaction time tests and graded as follows.

Grade	Reaction time in seconds	Tally	Frequency
A	0–		
B	0.2–		
C	0.4–		
D	0.6 – 0.8		

The following times were recorded:

0.1,   0.2,   0.3,   0.16,   0.23,   0.71,   0.19,   0.24,   0.3,   0.31,
0.35,   0.38,   0.31,   0.39,   0.62,   0.51,   0.44,   0.59,   0.4,   0.5.

(a) Complete the tally and grouped frequency distribution on a copy of the above table.

(b) After training, all pilots previously in grades *B*, *C* and *D* improved their times by 0.05 seconds.
How many pilots are now in the *C* grade range?   [SEG]

8. Natasha was asked to find an estimate of the mean height of adult women. To save time she found the mean height of her 10 aunts. Give **two** reasons why this sample may not give an accurate result.
                                                                            [NEAB]

9. A large ceramics company employing 3900 people wishes to undertake a survey to investigate their views on the introduction of a new performance related pay structure.

(a) Give a reason why the company might initially undertake a pilot survey.

(b) Explain how a simple random sample of 130 employees may be selected from the work force for the purpose of the main stage survey.

(c) The breakdown of employees in each section of the company in 1994 was as follows:

Section	Management	Clerical	Technical	Manual
No. of employees	65	200	1295	2340

Use this information to describe how a stratified random sample of 780 employees could be selected, showing the number to be chosen from each section.   [NEAB]

10. Amanda wants to choose a sample of 500 adults from the town where she lives.
She considers these methods of choosing her sample.

Method 1: Choose people shopping in the town centre on Saturday mornings.
Method 2: Choose names at random from the electoral register.
Method 3: Choose people living in the streets near her house.

Which method is **most** likely to produce an unbiased sample? Give a reason for your answer.   [NEAB]

11. A school needed planning permission to construct an all-weather playing surface with floodlights.
The school planned to distribute questionnaires in order to find the level of local support.

(a) State **one** advantage of using a sample to obtain the required information.

Questionnaires were given to the families of pupils in the school.

(b) Explain why this method of selecting the sample was unsuitable.
(c) Describe a better way of obtaining the sample of people to be given questionnaires.

The school wished to know the ages of the people completing the questionnaires.

(d) Write out a question suitable for obtaining this information, exactly as it should appear on the questionnaire.   [SEG]

12. (a) What is the main difference between obtaining information by census and by sampling?
(b) Why is it necessary to use sampling in a survey to find the lifetime of electric bulbs?
(c) Why is it usual to use sampling for an opinion poll?   [NEAB]

13. To improve sales it is necessary to find out what the public think of all kinds of products and services.
Questionnaires are often used to obtain this information.

(a) Write down **two** important points which should be remembered when designing a questionnaire.

Questionnaires are often sent by post.

(b) (i) Give **one** advantage of using the postal system.
   (ii) Give **one** disadvantage of using the postal system.

A food company intended to ask the following question in a questionnaire:

'How heavy are you?'

(c) (i) Why is this question unacceptable?
   (ii) Rewrite the question to make it acceptable.   [SEG]

14. Give a reason why questions A and B should be re-worded before being included in a questionnaire. Rewrite each one showing exactly how you would present it in a questionnaire.

Question A: How old are you?

Question B: The new supermarket seems to be a great success. Do you agree?   [NEAB]

15. A market research company is conducting a survey to find out whether most people had a holiday in Britain, elsewhere in Europe or in the rest of the world, last year. It also wants to know if they stayed in self-catering accommodation, hotels or went camping.

Design **two** questions that could be used in a questionnaire to find out all this information.   [NEAB]

16. A report in a medical journal says that people are taller in the morning than in the evening. Describe a simple statistical experiment to test this assertion.   [NEAB]

17. A dentist wishes to investigate the effectiveness of a new brand of toothpaste.
He chooses 50 boys and 50 girls at random from his patients.
The girls are given the new brand of toothpaste and the boys are given a different brand.
After 6 months the dentist compares the boys' and girls' teeth.

(a) Give **two** reasons why this is not a reliable experiment.
(b) Give **two** ways in which the experiment could be improved.   [NEAB]

18. A new brand of 'slim-fast' milk has been introduced for sale into a store. It is claimed that users will achieve significant weight loss after using the product for a period of seven consecutive days.

A statistical experiment is to be set up to test this claim.
The experiment will involve using 50 members each in both
experimental and control groups. These participants are to be
selected from the first 200 shoppers entering the store on a given day.

Explain briefly
(a) why a control group should be used in this case,
(b) how members of the control and experimental groups should
   be selected if paired comparisons are to be made,
(c) what procedures should be followed to ensure valid conclusions
   are reached from this experiment.   [NEAB]

19. Twenty workers were asked to fill in this questionnaire.

> Throw a coin.
>
> If it shows a HEAD, tick the YES box below.
>
> If it shows a TAIL, answer the question "DO YOU
> EVER SMOKE AT WORK?"
>
> YES ☐   NO ☐

(a) How many HEADS would you expect?
(b) If all the workers smoke at work, how many forms will have the
   YES box ticked?
(c) If no workers smoke at work, how many forms would you
   expect to have the YES box ticked?
(d) When the forms are returned, 16 have the YES box ticked.
   Estimate the number of workers who smoke at work.
(e) What is the advantage of using a questionnaire like this?
                              [NEAB]

20. One of the questions from a questionnaire used in a survey is shown.

> Are you tall?   Please tick   Yes ☐
>   No ☐

Out of 100 people who took part in this survey 20 ticked 'Yes' and
30 ticked 'No'.
There was a non-response of 50.
(a) What proportion of those that answered this question ticked 'Yes'?
(b) Give two reasons why the 'non-response' category cannot be
   ignored.
(c) Rewrite the question in a more suitable form.   [SEG]

21. The number of pupils in each Year Group of a school is given in the following table.

Year Group	7	8	9	10	11
Number of pupils	100	75	125	100	100

(a) Describe how a **simple random** sample of 80 pupils should be selected.
(b) Describe how a **stratified** sample of 80 pupils should be selected.
(c) Describe how a **quota** sample of 80 pupils should be selected.

[SEG]

22. A car rental firm intends to conduct a survey to help improve its service to the public. Before employing a consultancy firm to conduct the survey, the car rental firm decides to carry out its own small survey.

Arriving at the local station at 11 am, employees of the car rental firm spend an hour questioning travellers.

(a) Give **two** different reasons why this survey is unsatisfactory.
(b) The consultancy firm recommends either a systematic or a stratified sample.
   (i)   How might the **systematic** sample be taken?
   (ii) How might the **stratified** sample be taken?
   (iii) What is the main difference between these sampling methods?
(c) Write a question to find the age of the people likely to use the firm.

[SEG]

23. The number of workers on the three floors of a factory are shown.

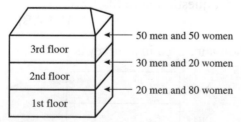

A different type of work is done on each floor.
The owner wants to ask 50 workers what they think of their working conditions.

(a) Give **one** advantage of taking a sample of the workforce to obtain this information.
(b) Explain why a **random** sample is considered to be an unsuitable way of selecting this sample.

It was decided to obtain a **stratified** sample according to the number of workers on each floor.

(c) Calculate the number of workers that should be questioned from each floor.

(d) How many women should be included from the 2nd floor to make the sample a fair representation of that floor?

(e) Explain how you would finally make a systematic selection of the women who would represent the 2nd floor.   [SEG]

24. A teacher wishes to select a representative sample of five pupils from a class of 18 boys and 12 girls. The class register is shown below. (B) indicates a boy and (G) indicates a girl.

Number	Name	Number	Name
1	C Baker (G)	16	N Jewell (G)
2	P Brown (B)	17	T Kennedy (B)
3	H Cant (G)	18	G Lomas (B)
4	H Curtis (B)	19	L Manors (G)
5	E Daly (G)	20	M Monet (B)
6	W Diamond (B)	21	K Nutt (B)
7	L Edwards (B)	22	F Oliver (B)
8	P Fletcher (G)	23	H Patel (B)
9	T Flynn (G)	24	D Peters (B)
10	W Ghani (B)	25	B Quarishi (B)
11	A Glade (G)	26	H Rayson (G)
12	B Hale (G)	27	C Rogers (B)
13	X Hatcher (B)	28	L Stonor (B)
14	F Isaacs (G)	29	P Taylor (B)
15	J Jacobs (B)	30	K Wood (G)

The teacher selects the five pupils by using the random numbers below. She starts at the top left of the random numbers and goes along the rows selecting two-digit numbers. If the random number is a register number the corresponding pupil is accepted. Otherwise the number is ignored. For example, the first two pupils selected are C Rogers and M Monet.

27 74 87 94 57 38 94 20 29 70 53 33 24 95 49
49 74 25 44 52 92 51 86 10 02 38 66 59 18 23

(a) Write down the names of the other three pupils selected.

(b) Why is the choice of pupils **not** satisfactory in this case?

(c) Suggest a better way to select a representative sample of five pupils.   [NEAB]

25. In a street with 30 houses, it was decided to ask a sample of five residents if they approve of the Council's proposed introduction of speed restriction humps in the road.

The map shows their location.

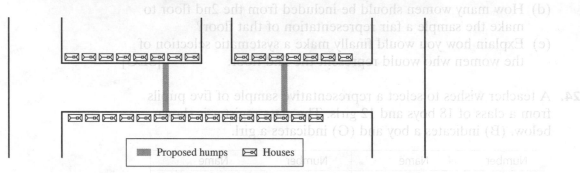

(a) Explain how you would choose a random sample of five for this enquiry.

It was suggested that a systematically chosen sample could be taken.

(b) Explain how you would select such a sample.
(c) Which of these two sampling methods would you choose?
Give a reason for your answer.   [SEG]

2 Interpretation of data

*A judicious man looks at statistics, not to get knowledge
but to save himself from having ignorance foisted on him*

Carlyle

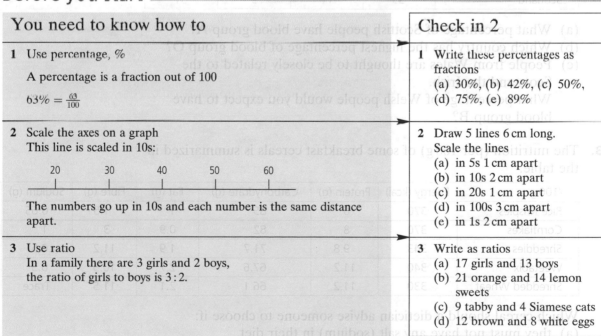

> This advertising is
> misleading because the
> 'extra free' part of the label
> represents more than 15%

This unit will show you how to

+ Find information from tables
+ Explain why diagrams can be misleading

Before you start

You need to know how to	Check in 2
1 Use percentage, % A percentage is a fraction out of 100 $63\% = \frac{63}{100}$	**1** Write these percentages as fractions (a) 30%, (b) 42%, (c) 50%, (d) 75%, (e) 89%
2 Scale the axes on a graph This line is scaled in 10s: 20 30 40 50 60 The numbers go up in 10s and each number is the same distance apart.	**2** Draw 5 lines 6 cm long. Scale the lines (a) in 5s 1 cm apart (b) in 10s 2 cm apart (c) in 20s 1 cm apart (d) in 100s 3 cm apart (e) in 1s 2 cm apart
3 Use ratio In a family there are 3 girls and 2 boys, the ratio of girls to boys is 3 : 2.	**3** Write as ratios (a) 17 girls and 13 boys (b) 21 orange and 14 lemon sweets (c) 9 tabby and 4 Siamese cats (d) 12 brown and 8 white eggs

2.1 Reading tables ▬▬▬▬▬

Data is often summarized in tables. This makes comparisons easier.
You need to be able to read information from these tables.

Exercise 2A _____

1. The composition of British coinage is

	Copper	Nickel	Zinc	Tin
1p, 2p	97%	–	2.5%	0.5%
5p, 10p, 50p	75%	25%	–	–
20p	84%	16%	–	–
£1	70%	5.5%	24.5%	–

   (a) Which coin(s) contain the least amount of copper?
   (b) Which coin(s) contains no zinc? What type of coinage are they?

2. The table shows the distribution of blood groups amongst the
   population of some countries in the mid-1980s.

% Blood group / Country	A	B	AB	O
England	42	8	3	47
Ireland	32	11	3	54
Scotland	35	11	3	51

   (a) What percentage of Scottish people have blood group A?
   (b) Which country has the highest percentage of blood group O?
   (c) People from Wales are thought to be closely related to the
       Scots and the Irish.
       What percentage of Welsh people would you expect to have
       blood group B?

3. The nutrition (per 100 g) of some breakfast cereals is summarized in
   the table.

/100 g	Energy (kcal)	Protein (g)	Carbohydrate (g)	Fat (g)	Fibre (g)	Sodium (g)
Rice Krispies	370	6	85	1	1.5	0.65
Cornflakes	370	8	82	0.9	3	1
Shreddies	343	9.8	71.7	1.9	11.2	0.5
Weetabix	340	11.2	67.6	2.7	10.5	0.3
Shredded Wheat	330	11.2	66.1	2.1	11.5	Trace

Which cereal should a dietician advise someone to choose if:
(a) they must not have any salt (sodium) in their diet,
(b) they should not have any fibre in their diet,
(c) they need to have a high fibre, low fat diet?

4. Rachael and Reuben changed schools. They were told that they had to attend at least two exercise classes. The table outlines the amount of benefit they may enjoy from some of the exercises.

Exercise	Amount of benefit		
	Strength	Stamina	Suppleness
Badminton	**	**	***
Cricket	*	*	**
Football	***	***	***
Gymnastics	***	**	****
Rowing	****	****	**
Squash	**	***	***
Swimming (hard)	****	****	****
Tennis	**	**	***
Weightlifting	****	*	*
Yoga	*	*	****

Key:
* no real effect
** beneficial
*** very good
**** excellent effect

(a) Which exercises should Rachael choose if she wants to become more supple?

(b) Which exercises should Reuben choose if he wants to increase his stamina and become stronger?

(c) Which exercises would provide the same amount of benefit?

(d) Give a reason for someone to choose Football instead of Squash.

2.2 Misleading statements

It is often not what is written, but what has been left out that is important.

Example

Critically analyse the statement:

'Sales increased by £3 million over the last year.'

This appears to mean that the company had a very successful year but ...

✦ If in the previous year sales were £10 000 it would represent a huge increase.

✦ If in the previous year sales were £300 million it would be a small increase.

The 'sales' only tell you how much is sold; you also need to know what happened to the profits.

● **Example**

Critically analyse the statement:

> 'Eight out of ten owners who expressed a preference said their cats preferred Catto.'

This does not mean that every owner asked expressed a preference, and it does not say how many of those asked actually expressed a preference.

You also need to consider:

✦ Where the researcher found the cat owners to ask.
✦ Whether the owners were found randomly or were they buying Catto tins in a supermarket?
✦ Whether the researchers were brandishing discount vouchers to be given out if owners said their cat preferred Catto.

> It could be that 1000 people were asked and only 100 expressed a preference of which 80 prefer Catto.

| A statement concerning a simple set of statistics rarely proves anything. |

Exercise 2B

1. A survey of members of a gardening club found that 78% of its members preferred home grown vegetables.
   Explain why this result is not necessarily representative of all people.

2. The following headline was seen in a newspaper.

37% of British children achieve less than average marks in science exams

Shock! Horror!

Explain why this news is not shocking.

3. Criticize the following statements.
   (a) 90% of people questioned approved of the site for the new maggot farm.
   (b) Air accident fatalities increased from 1927 to 1987, so it is more dangerous to fly in 1987 than it was in 1927.
   (c) In a survey of 5000 people the bread-making machine 'Dougood' was the preferred choice for home baking.

2.3   Pictorial misrepresentation

False zeros

> The most common way to present correct, but misleading, data is to use a false origin.

⬤ **Example**

Critically comment on these bar charts which represent the number of kettles sold over two years at a local store.

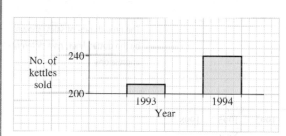

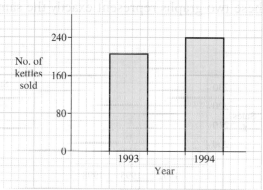

This chart gives the impression that sales were four times greater in 1994 compared with 1993. However, the sales do not start at zero.

When the whole chart is drawn you can see that although sales have increased they have not even doubled.

Profits for the store, over three years, are shown on this line diagram.

The impression given is that there was a sharp rise in profits over the years.

However, the origin has been omitted.

It is impossible to comment on the graph unless you can work out the scale on the vertical axis.

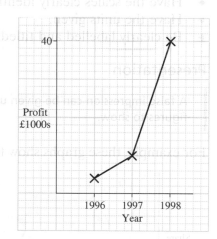

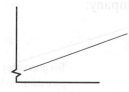

If you draw a graph with a scale that does not begin at zero then it should be identified with a squiggle on the axis as shown.

Scales and labelling

> You can change the impression given by a graph by varying the scale or omitting a scale.

For example: these graphs have been drawn with no vertical scale:

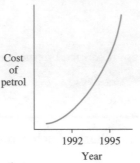

Both graphs may have been drawn to the same scale. You cannot decide if the cost of petrol has increased slowly, as in the first graph; or quickly, as in the second graph.

These two graphs represent exactly the same data:

The growth rate of sales, shown by the dotted lines, give different impressions.

> Omission or misuse of a scale is misleading.

All diagrams should:

- ✦ Have the scales clearly identified.
- ✦ Have the units given.
- ✦ Be clearly labelled and titled.

Presentation

> A false impression can be given using a graph by carefully selecting which figures to show.

For example, these graphs show the change in share price for a company:

This graph stops before the share price started to fall. This makes the company look more attractive to investors.

You can also see that using shadows or thick lines to join points makes it difficult to read graphs.

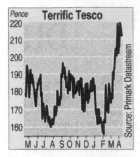

The Mail, April 15th, 2000

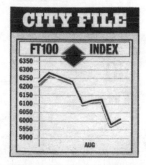

The Express, Thursday August 12th, 1999

Exercise 2C

1. The following graphs all appeared in national newspapers. Criticize their presentation.

(a)

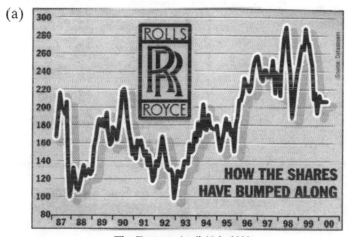

The Express, April 11th, 2000

(b)

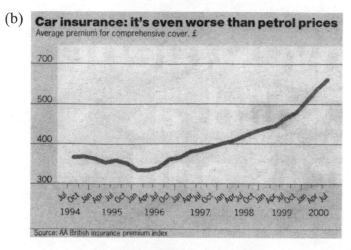

The Guardian, August 26th, 2000

(c)

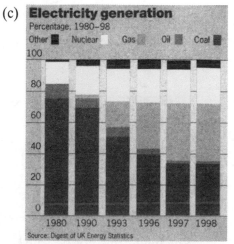

The Guardian, May 23rd, 2000

(d)

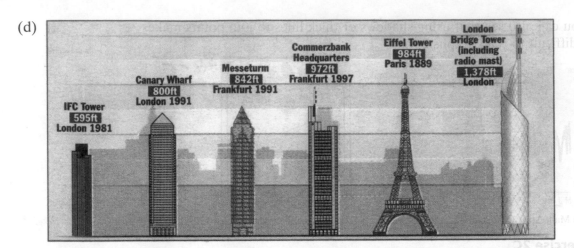

The Express, April 11th, 2000

2. This graph appears to show a big increase in sales of Brisk:

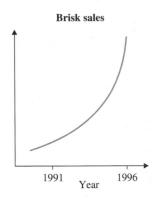

Are you convinced of the success of Brisk?
Give a reason for your answer.

3. This graph implies that price rises in the cost of CDs are slowing down.

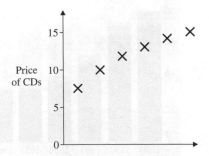

Why might this not be true?

4. Which of these two businesses is growing faster?
Explain how you found your answer.

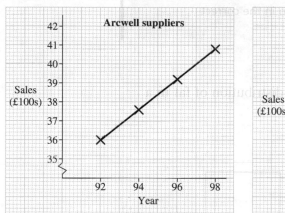

5. The following diagrams appeared in a financial report to represent sales and profits over two years.

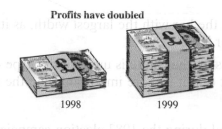

Profits have doubled

1998   1999

(a) Explain why the sales diagram is misleading.
(b) Explain why the profit diagram is misleading.
(c) What extra data is needed before you could redraw the sales diagram to make it fair?

6. Sales of calculators in a school shop are given in the table below.

Term	Sales
Autumn	74
Spring	62
Summer	86

Draw two line graphs to represent this information with a vertical scale:

(a) from 60 to 90 with 2 cm representing 10 calculators,
(b) from 0 to 90 with 2 cm representing 30 calculators.

Comment on the different impressions given by the two graphs.

2.4   Misleading diagrams

Diagrams can be misleading if they are not drawn in the correct proportions.

● **Example**

What is wrong with this bar chart showing the distribution of fat in a tin of salmon?

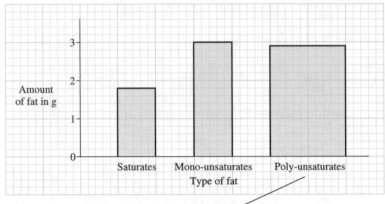

Your eye is drawn to the bar with the largest width, as its area is greater than the other bars.

This gives the impression that there is more of this type of fat than the others; or at least that it is more important than the others.

This diagram was used during the 1983 election campaign:

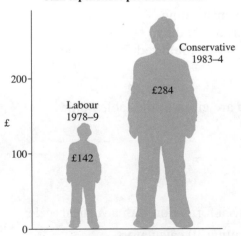

Cash expenditure per head on NHS

Your eye is drawn to the larger diagram. Most people would tend to ignore the figures, which show that expenditure had doubled, and concentrate on the visual impact of one diagram compared with the other, which is four times larger.

The heights of the diagrams are in proportion to the amounts spent, but the areas of the diagrams are not in proportion.

> When you draw diagrams for pictorial comparisons you should take care not to misuse length, area and volume.

Exercise 2D

1. The diagram represents sales of cakes in a bakery.
   (a) What is the ratio of:
       (i) their heights,
       (ii) their areas?
   (b) Does the diagram give a fair representation of the cake sales? Give a reason for your answer.

2. The diagram shows information about the sales of razor blades in 1989.

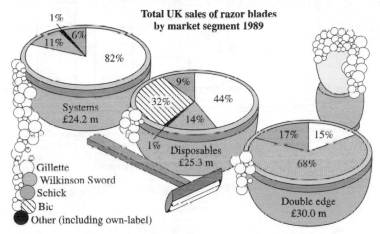

   How could you change the pie charts to show a fairer representation of the different types of razor blades sold?

3. The diagram shows the reported non-collections of refuse as a percentage of the total number of households.

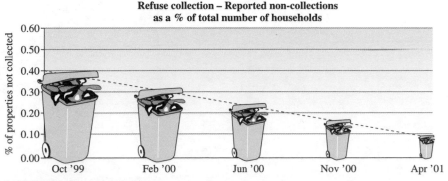

   Critically comment on the diagram.

Summary

You should now be able to	Check out 2
1 Find information from tables.	1 Which of these drinks would give you (a) the least carbohydrate, (b) the most energy?
2 Explain why diagrams can be misleading.	2 Why is this diagram misleading?

1 Which of these drinks would give you (a) the least carbohydrate, (b) the most energy?

Per 100 ml diluted	Energy	Protein	Carbohydrate
Citrus	29	< 0.1	4.9
Exotic fruit	9	< 0.1	2
Garden berry	30	Trace	7.4

2 Why is this diagram misleading?

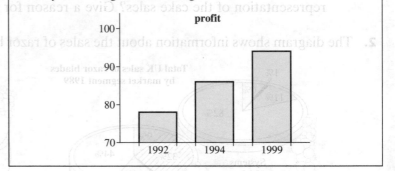

profit

Revision Exercise 2

1. The Government produced the following accounts for the year
1995–96.

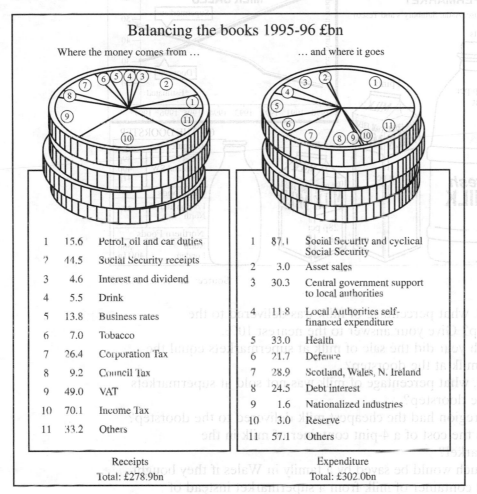

Balancing the books 1995-96 £bn

Where the money comes from …   … and where it goes

	Receipts	
1	15.6	Petrol, oil and car duties
2	44.5	Social Security receipts
3	4.6	Interest and dividend
4	5.5	Drink
5	13.8	Business rates
6	7.0	Tobacco
7	26.4	Corporation Tax
8	9.2	Council Tax
9	49.0	VAT
10	70.1	Income Tax
11	33.2	Others

Receipts
Total: £278.9bn

	Expenditure	
1	87.1	Social Security and cyclical Social Security
2	3.0	Asset sales
3	30.3	Central government support to local authorities
4	11.8	Local Authorities self-financed expenditure
5	33.0	Health
6	21.7	Defence
7	28.9	Scotland, Wales, N. Ireland
8	24.5	Debt interest
9	1.6	Nationalized industries
10	3.0	Reserve
11	57.1	Others

Expenditure
Total: £302.0bn

(a) Where does the largest amount of money come from?
(b) How much did the Government spend on Defence?
(c) How much did the Government need to borrow to meet the
expenditure total of £302.0bn?   [SEG]

2. The following headline was taken from a newspaper.

> **SHOCKING NEWS.** 40% OF CHILDREN IN BRITISH
> SCHOOLS GET LESS THAN THE AVERAGE NUMBER
> OF MARKS FOR MATHEMATICS

Explain in your own words why this is **not** shocking news.   [NEAB]

3. The diagram shows the results of a survey about the sale of milk.

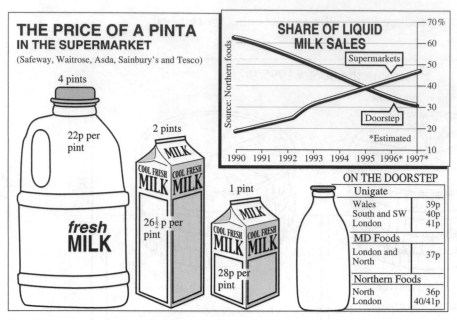

Source: *The Times* 24.3.95

(a) In 1990, what percentage of milk was delivered to the doorstep? Give your answer to the nearest 10%.

(b) In which year did the sale of milk at supermarkets equal the sale of milk at the doorstep?

(c) In 1993, what percentage of milk was not sold at supermarkets or at the doorstep?

(d) Which region had the cheapest milk delivered to the doorstep?

(e) What is the cost of a 4-pint container of milk in the supermarket?

(f) How much would be saved by a family in Wales if they bought a 4-pint container of milk from a supermarket instead of having 4 pints delivered to the doorstep?   [NEAB]

4. The following diagram was produced to give information about cereal production in Morocco.

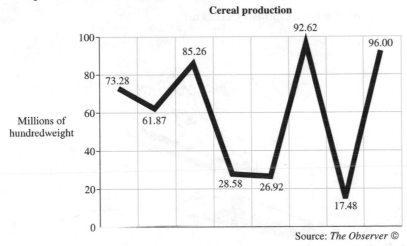

Cereal production

Source: *The Observer* ©

Make **three** comments saying why the diagram could be misleading.

[SEG]

5. The table below shows the results of a survey about the three basic skills.

BASIC SKILLS

2875 people were asked about any difficulties they had experienced with basic skills since leaving school.

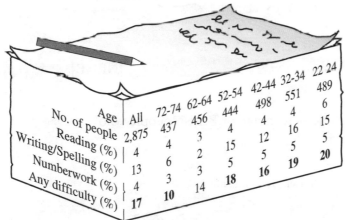

Age	All	72-74	62-64	52-54	42-44	32-34	22 24
No. of people	2,875	437	456	444	498	551	489
Reading (%)	4	4	3	4	4	4	6
Writing/Spelling (%)	4	4	2	15	12	16	15
Numberwork (%)	13	6	5	5	5	5	5
Any difficulty (%)	4	3	3	18	16	19	20
	17	10	14				

Source: *ALBSU* 1994

(a) What percentage of 22–24 year olds found reading difficult?
(b) 16% of which age group had difficulty with writing/spelling?
(c) Some people had difficulty with more than one basic skill. How can you tell this from the table?
(d) There is a mistake in the 62–64 column. How can you tell?

[NEAB]

6. Criticize the diagram below.

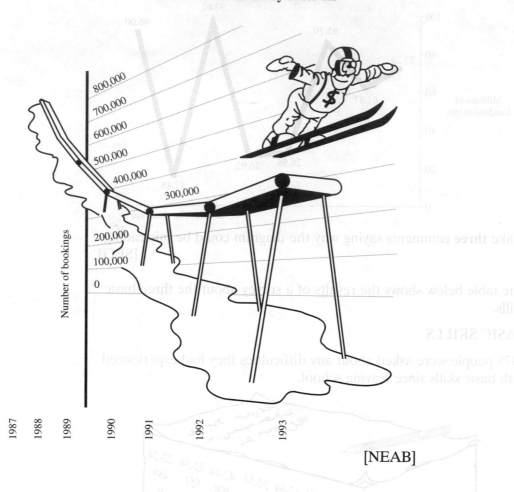

Winter resort industry takes off

[NEAB]

7. The diagram below shows the percentage of British people who own their own house.

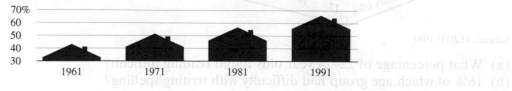

Growth of owner occupation as a % of all British householders

State **two** ways in which the diagram is misleading.   [NEAB]

8. The diagram shows the trade in millions of dollars ($m) between different areas in the world.

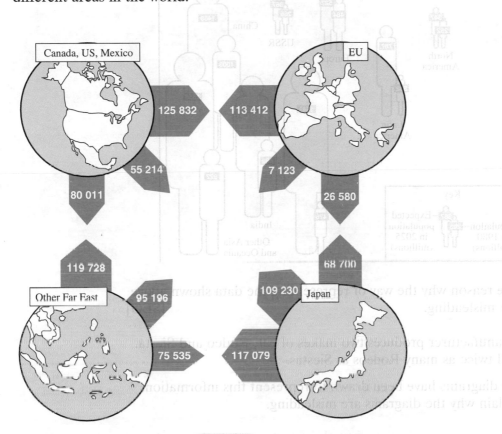

The diagram shows that Japan exported 68 700 $m of goods to the EU and imported goods to the value of 26 580 $m from the EU.

(a) Find the value of imports received by Canada, US and Mexico from Japan.
(b) Work out the difference in value of the trade between the EU and 'Other Far East' countries.
(c) Calculate the total difference in value between the exports and imports for trading between the EU and the other areas in the diagram.   [SEG]

9.

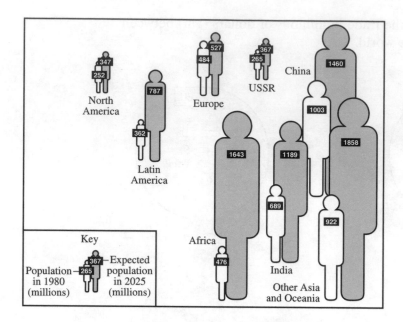

Give **one** reason why the way of representing the data shown above could be misleading.   [SEG]

10. A car manufacturer produces two makes of car, Rodeo and Siesta. They sell twice as many Rodeos as Siestas.

(a) The diagrams have been drawn to represent this information. Explain why the diagrams are misleading.

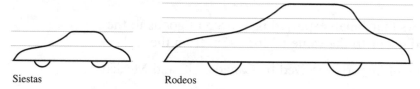

Siestas   Rodeos

The manufacturer decides to make models of the two cars in proportion to their sales.
The model of the Rodeo will be twice the volume of the model of the Siesta.

The model of the Siesta will be 4 cm long.

(b) Calculate the length of the model of the Rodeo.   [SEG]

3 Tabulation and representation

Diagrams are often used to display data:

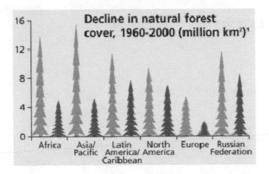

This unit will show you how to

+ Draw diagrams to represent discrete data
+ Draw diagrams to represent continuous data
+ Decide on the most appropriate method of display
+ Interpret the shape of a distribution

Before you start

You need to know how to	Check in 3
1 Read scales	1 2 2.5 ↑ 3 (a) Estimate the value marked by an arrow. (b) Copy the diagram and mark the position of (i) 2.2 (ii) 2.65
2 Plot coordinates	2 Plot the following coordinates A $(1, 2)$; B $(3, {}^-2)$; C $({}^-1, 0)$ D $({}^-3, {}^-2)$ and E $(1, {}^-2)$. Join ABCDE with straight lines.
3 Use a protractor	3 Draw angles of (a) $65°$ (b) $157°$ (c) $205°$
4 Divide decimals	4 Calculate (a) $8 \div 0.4$ (b) $0.6 \div 0.5$ (c) $0.6 \div 0.03$ (d) $0.008 \div 0.02$
5 Calculate using percentages	5 Calculate (a) 20% of 75 (b) 18% of 250 (c) Express $\frac{50}{80}$ as a percentage.

3.1 Displaying discrete data

p269

p270

There are many ways of displaying discrete data.

You will have met some of them before in your maths lessons.

In GCSE Statistics you need to know about these ways of displaying discrete data:

- ✦ Pictograms
- ✦ Bar charts, including multiple and composite bar charts
- ✦ Stem and leaf diagrams
- ✦ Pie charts
- ✦ Line graphs
- ✦ Cumulative frequency step polygons
- ✦ Shading maps

3.2 Pictograms

> A pictogram uses pictures or symbols to represent data.

Example

The table shows the number of cars sold by a garage in a three-month period:

Month	May	June	July
Number of cars	40	32	28

Show this data on a pictogram.

Use a key so that one car represents 10 sales.

Then to represent 2 sales, you draw part of the car:

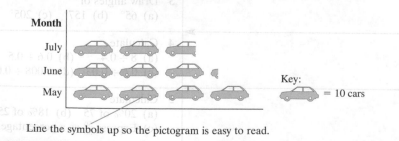

Line the symbols up so the pictogram is easy to read.

Remember to include a key so other people can understand your graph.

Exercise 3A

1. The table shows the populations of five towns (to the nearest hundred).

Town	Firston	Seconton	Thirton	Fourton	Fifton
Population	5500	2000	1500	3300	4000

(a) Draw a pictogram to represent this data.
Use a key so that one symbol equals 500 people.

(b) Which of these populations would be difficult to interpret accurately?

2. The pictogram shows the number of cars manufactured by four rival companies in a particular year.

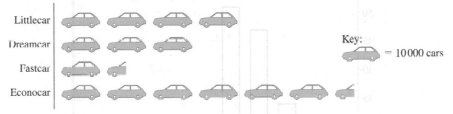

(a) How many cars did each of the companies make?

(b) A fifth company, Spacecar, claims to have made 50% more than Econocar in the same year. Illustrate Spacecar's sales as they would appear on the pictogram.

3. The number of books in the libraries of 4 university departments are

English	History	Mathematics	Statistics
14,000	13,000	10,500	500

(a) Draw a pictogram to represent this data. Use one symbol to represent 2,000 books.

(b) A statistician decides to combine two groups together. Explain which groups could be combined and what is the benefit of combining them.

3.3 Bar charts

p269

p270

A bar chart uses bars to represent data.

A bar chart can be drawn horizontally or vertically.

When you draw a bar chart you should:

✦ Choose an appropriate scale.

✦ Ensure the bars are the same width.

✦ Always label your axes.

Example

Draw a bar chart to illustrate this data showing the sales of a particular ice-cream over a 6-month period.

Month	April	May	June	July	August	Sept
Sales	11	28	34	47	51	22

Choose an appropriate scale:

The highest value is 51 and the lowest is 11, so a good scale is going up in 5s.

Vertical bar chart:

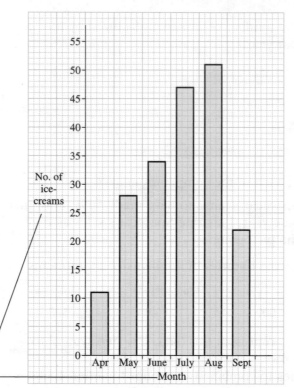

Always label your axes.

Horizontal bar chart:

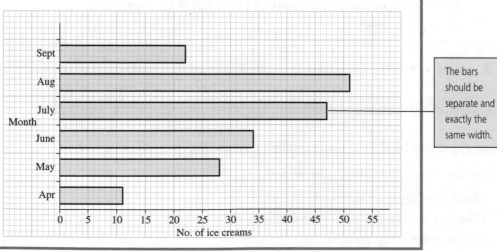

The bars should be separate and exactly the same width.

Vertical line graphs

> A vertical line graph is like a bar chart except it uses lines instead of bars.

For example this is the vertical line graph for the data in the example on the previous page:

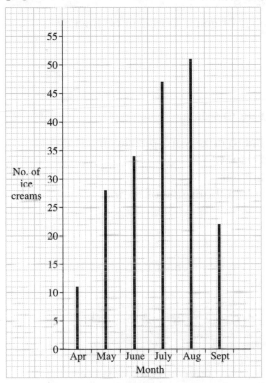

Exercise 3B

1. The following bar chart shows the results of a survey in which 100 people were asked how they travelled to school each day.

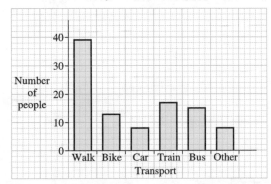

  (a) What is the preferred method of transport?
  (b) How many more pupils walk than travel by bike?
  (c) Draw a pictogram to illustrate this data.

2. The following table shows the number of children in 50 households surveyed.

Number of children	0	1	2	3	4	5
Number of households	12	13	15	6	3	1

Illustrate this data with a vertical line graph.

3. The following table shows the number of goals scored in 60 hockey matches.

Number of goals	0	1	2	3	4	5	6
Number of matches	2	6	14	18	10	6	4

Draw a vertical line graph to illustrate this data.

4. Collect data on the length of answers to crossword puzzles in (i) a broadsheet paper and (ii) a tabloid paper.
   Use the number of letters in each answer.
   (a) Use a vertical line graph to illustrate both sets of data.
   (b) Compare the two sets of data.

Multiple and composite bar charts

Another way of showing two sets of data on a bar chart is to show them side by side or as components of a whole.

> A bar chart showing two or more sets of data side by side is a multiple bar chart.

> A chart showing two or more sets of data as components of a whole is a composite bar chart.

Example

Here is a multiple bar chart showing how many students study each subject in each year group.

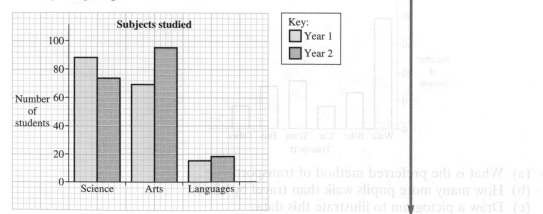

(a) How many more Arts students are there in year 2 than in year 1?
(b) Draw a composite bar chart to show the data for the choices of students each year.
(c) Which chart would you use to compare the total number of students in each year?
(d) Which chart best displays the data? Give a reason for your answer.

(a) In year 1 there are 69 Arts students and in year 2 there are 95. So there are 26 more in year 2.
(b) The chart shows that in year 1, 88 students studied Science, 69 studied Arts and 15 studied Languages which gives a total of 172 year 1 students.
Similarly in year 2, 74 studied Science, 95 Arts and 18 Languages giving a total of 187 year 2 students.

The composite bar chart looks like this:

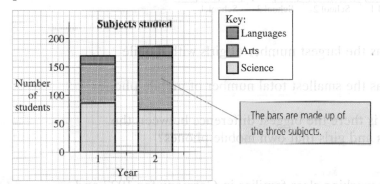

(c) The composite bar chart is easier to use as it shows the total number of students each year.
(d) The answer depends on what you want to use the data for. If you want to compare take-up rates for each subject then the multiple bar chart is the best to use. It is also easier to read.
However, if you want to compare totals, it is much easier to use the composite bar chart as you can see the total easily.

Exercise 3C _____

1. The following table shows the number of students at Lowbrow University over three consecutive years.

Year	1998	1999	2000
Number of students aged under 25	3000	2500	2500
Number of students aged 25 and over	500	1000	1500

(a) Draw a composite bar chart showing this data.
(b) Describe the changes in the total number of students and in the age proportions.

2. The multiple bar chart below shows the number of mobile phone
users in four secondary schools.

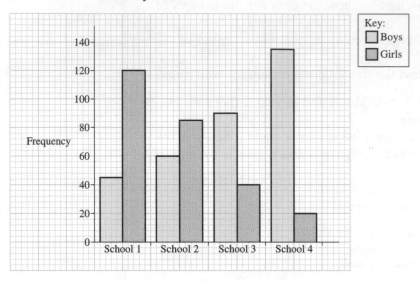

(a) Which school has the largest number of girls with mobile
    phones?
(b) Which school has the smallest total number of mobile phone
    users?
(c) In which school is there the biggest difference between the
    numbers of boys and girls that own mobile phones?

3. The consumption by working class families in Germany for 1927 and
1937 are given in the following table.

	Bread	Meat	Potatoes	Other foods
1927	318 kg	155 kg	499 kg	224 kg
1937	347 kg	131 kg	520 kg	211 kg

Illustrate the data using
(a) A multiple bar chart.
(b) A composite bar chart.

4. The population (in millions) of California by ethnic group is given in
the following table.

	White	Hispanic	Black	Other
1980	16.8	3.8	1.8	1.3
1990	17.9	6.4	1.8	3.5

Illustrate the data using a variety of charts of your choice.
Explain the advantages of each of your chosen charts.

3.4 Pie charts

A pie chart is useful when you want to show proportions – you show each category as a slice of a pie (a circle). It is particularly useful when there are only a few categories.

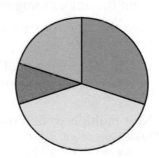

> A pie chart uses a circle to display data.

To draw a pie chart you:

✦ work out the fraction of the total each category uses

✦ then work out how many degrees this represents, that is the fraction of 360°.

● Example

Draw a pie chart to illustrate this data for Richman Garage:

Month	Sales of Mercedes
May	50
June	40
July	30

The total sales for the garage is $50 + 40 + 30 = 120$.

Draw a table with headings like this:

Month	Sales	Fraction	Angle
May	50	$\frac{50}{120}$	$\frac{50}{120} \times 360° = 150°$
June	40	$\frac{40}{120}$	$\frac{40}{120} \times 360° = 120°$
July	30	$\frac{30}{120}$	$\frac{30}{120} \times 360° = \ 90°$
Total	120	$\frac{120}{120}$	$360°$

Now draw the pie chart:

1 Start with a line from the centre.

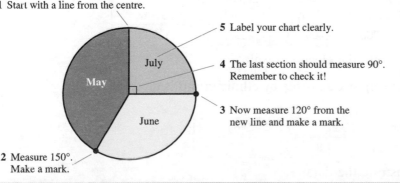

5 Label your chart clearly.

4 The last section should measure 90°. Remember to check it!

3 Now measure 120° from the new line and make a mark.

2 Measure 150°. Make a mark.

> Hint: Make sure your total is 360° otherwise you have made a mistake.

> Make sure you can use a protractor – practice drawing this chart before starting the exercise!

If you don't like working with fractions, here is an alternative way to work out the angles using proportions:

✦ Find the total for your set of data.
   In the example it is 120.

✦ Divide 360° by the total.
   360° will display 120 items so each item uses:
   360° ÷ 120 = 3°

> If 360 ÷ total gives a decimal number, be careful when you round that the total is 360°.

✦ Now multiply to find the angle for each category:
   50 sales uses 50 × 3° = 150°
   40 sales uses 40 × 3° = 120°
   30 sales uses 30 × 3° = 90°

✦ Draw the pie chart as before. Remember to check the total is 360° before you start!

Exercise 3D _____

1. The pie chart shows the results of a survey in which 120 randomly chosen people were asked what type of food they prefer.
   (a) How many people preferred Indian food?
   (b) (i) What is the angle representing Italian food?
       (ii) How many people prefer Italian food?

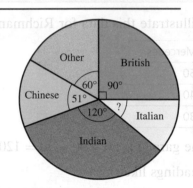

> Hint: Write the number of degrees for each category as a fraction of 360. This gives the fraction of the total.

2. The table shows the holiday destinations of 100 people.

UK	Europe	Asia	America	Africa	Australia	Other
29	33	10	5	7	6	10

Draw a pie chart to illustrate this data.

3. The table gives the number of votes cast in the German election of 1930.

Party	Social Democratic	Nazi Party	Communist Party	Centre Party
Number of votes	8,577,000	6,409,000	4,592,000	4,127,000

Draw a pie chart to illustrate the election result.

4. The population (in millions) of California by ethnic group is given in the following table.

White	Hispanic	Black	Other
17.9	6.4	1.8	3.5

Draw a pie chart to illustrate the data.

3.5 Comparative pie charts

You can use pie charts to display two or more sets of data. To do this accurately you must ensure that the areas of the circles are in the same proportion as the totals displayed on each chart.

Example

In 2000, 1000 people chose Fun Package Tours as their tour operator.

In 2001, their reputation grew and 20% more people chose them as their operator.

Fun Package Tours would like to show this data in their brochure using pie charts.

They choose a radius of 5 cm for the 2000 pie chart.

What radius should they use for the 2001 pie chart?

You need to compare the areas of the charts:

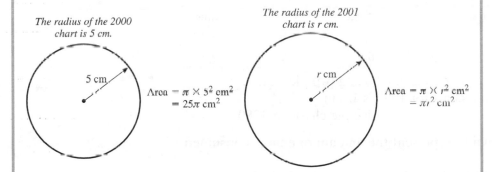

The radius of the 2000 chart is 5 cm.

The radius of the 2001 chart is r cm.

5 cm

Area $= \pi \times 5^2$ cm^2
$= 25\pi$ cm^2

r cm

Area $= \pi \times r^2$ cm^2
$= \pi r^2$ cm^2

The area of the 2001 chart should be 20% more than the area of the 2000 chart, or 120% of the area of the 2000 chart.

So $\pi r^2 = 120\% \times 25\pi$

Hence $r^2 = 120\% \times 25$

And $r = \sqrt{1.2 \times 25}$

So the radius of the 2001 chart should be 5.5 cm (to 1 d.p.).

Note that many students make the mistake of increasing the radius rather than the area. In this case, a 20% increase of the radius of 5 cm would be a radius of 6 cm. This radius would give an area of 36π cm^2. From Unit 2 you can see this is misleading.

To compare data using pie charts the areas of the circles must be in the proportion of the totals.

Exercise 3E _____

1. A college has 1000 pupils and another college has 2000 pupils.
   A student draws comparative pie charts to represent the number of
   students studying particular subjects at each college.
   He draws one with a diameter of 10 cm and the other with a
   diameter of 20 cm.
   (a) Explain why this is wrong.
   (b) Calculate the correct radius of the larger pie chart.

2. The following table gives the number of votes cast in the German
   elections of 1930 and 1932.

Party	Social Democratic	Nazi Party	Communist Party	Centre Party
1930	8,577,000	6,409,000	4,592,000	4,127,000
1932	7,959,000	13,745,000	5,283,000	4,589,000

   The above data is to be represented using two pie charts.
   The radius of the pie chart for 1930 is 4 cm.
   Calculate the radius of the comparative pie chart for 1932.

3. The population (in millions) of California by ethnic group is given in
   the following table.

	White	Hispanic	Black	Other
1980	16.8	3.8	1.8	1.3
1990	17.9	6.4	1.8	3.5

   The above data is to be represented using two pie charts.
   The radius of the pie chart for 1990 is 10 cm.
   Calculate the radius of the comparative pie chart for 1980.

4. Pie charts are drawn to represent the amount of energy consumed
   in 1970 and 1990.

	1970	1990
Radius	8 cm	10.2 cm
Energy consumed	26.3 million units	x

   Calculate the energy consumed in 1990.

3.6 Cumulative frequency step polygons ▬▬▬

You use cumulative frequency for a set of data to show how the data
grows, or accumulates.

Before you can draw any graphs you must use a table to work out the
cumulative frequencies.

> Cumulative frequency is the total frequency up to and including a
> particular data point.

Example

This data shows the length of words used in the solution to a crossword:

No. of letters	2	3	4	5	6	7	8	9	10	11
Frequency	0	0	12	9	8	8	13	4	0	5

(a) Make a table to show the cumulative frequencies for the data.
(b) Use your table to draw a cumulative frequency step polygon.

(a) The cumulative frequencies are:

No. of letters	3	4	5	6	7	8	9	10	11
Frequency	0	12	9	8	8	13	4	0	5
Cumulative frequency	0	12	12 + 9 — 21	21 + 8 — 29	29 + 8 = 37	37 + 13 = 50	50 + 4 = 54	54 + 0 = 54	54 + 5 = 59

(b) Plot the points:

(3, 0)   (4, 12)   (5, 21)   (6, 29)   (7, 37)   (8, 50)   (9, 54)   (10, 54)   (11, 59)

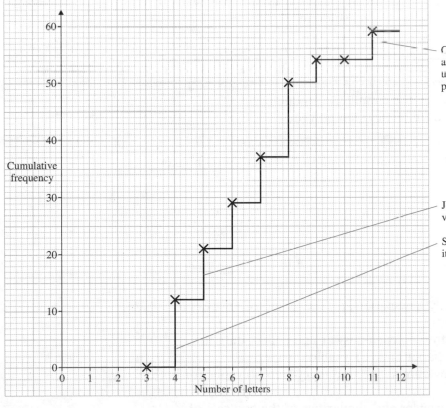

Continue to plot points and join them in steps until you reach the last point.

Join the points in vertical steps.

Start at (3, 0) and join it to (4, 12) in a step.

Note: that you don't join up the points diagonally. That is because the data is discrete and so can only take the values 3, 4, 5 etc. You only join the points diagonally when you are displaying continuous data, as you will see on page 72.

> In a cumulative frequency step polygon you plot the total frequency up to and including the current value. You join the plotted points together in steps.

Exercise 3F

1. The data shows the results of a survey in which 50 randomly chosen adults were asked how many times they had visited the cinema in the past year.

Number of visits	0	1	2	3	4	5	6
Frequency	6	13	17	7	4	2	1

(a) Make a table to show the cumulative frequencies for the data.
(b) Use your table to draw a cumulative frequency step polygon.

2. The data shows the number of GCSE passes at grades A–C gained by a sample of 100 first year A-level students.

Number of GCSE passes	4	5	6	7	8	9	10
Frequency	9	12	18	21	17	15	8

(a) Make a table to show the cumulative frequencies.
(b) Draw a cumulative frequency step polygon.

3. The following table shows the number of goals scored in 60 hockey matches.

Number of goals	0	1	2	3	4	5	6
Number of matches	2	6	14	18	10	6	4

Draw a cumulative frequency step polygon of the data.

4. Collect data, using daily newspapers, on the length of answers to crossword puzzles.
Use the number of letters in each answer and collect data for both (i) a broadsheet paper and (ii) a tabloid paper.
On the same axes draw cumulative frequency step polygons for both sets of data.
Comment on the differences and similarities of the two sets of data.

3.7 Choropleth maps

Choropleth maps are used extensively in geography.

They show, using shading, how quantities change.

Here are two examples:

Note: you must include a key.

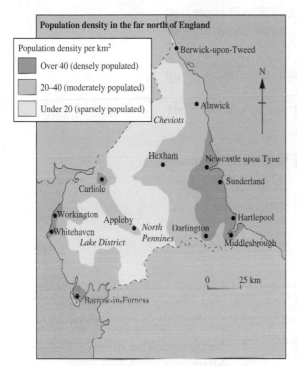

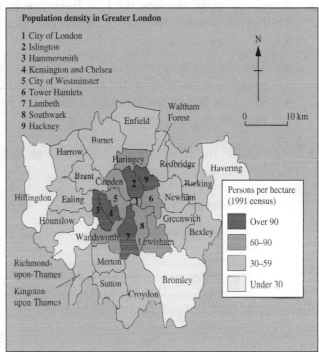

Notice that the denser the population, the darker the shade used.

Exercise 3G

In this exercise you will complete a choropleth map to show the population density for the countries in South America.
(Population density means the average number of people living in a square kilometre.)

Country	Population density
Argentina	10
Brazil	14
Bolivia	5
Chile	15
Colombia	24
Ecuador	29
Guyana	4
Paraguay	7
Peru	14
Suriname	2
Uruguay	16
Venezuela	15
French Guiana	1

1. First, find the highest and lowest values in the table. The country with the highest population density is _____ , with 29 people per square kilometre.

   The country with the lowest value is _____ , with 1 person per square kilometre.

2. Now copy and complete the key. Each box in the key will represent a range of population density. Most choropleth maps use from 4 to 6 boxes. You should use 6 in this case.

   ✦ The key has been started below. Notice how the numbers go up in equal steps. Fill in the missing numbers.
   ✦ Fill each box of the key with a different pattern or colour. Start with the lightest one at the bottom.

3. Now shade the rows of the table to match your key. The first group has been done for you!

   A choropleth map to show

   _____

   5. Give the map a title, by finishing of this sentence.

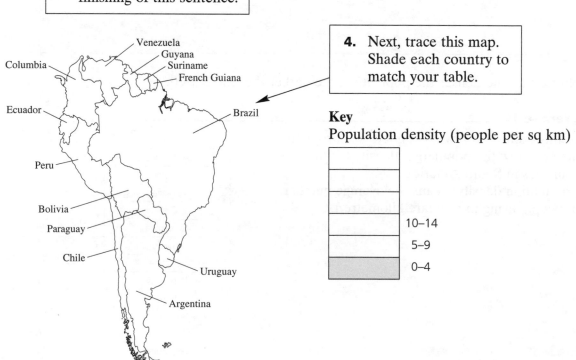

   4. Next, trace this map. Shade each country to match your table.

   Key
   Population density (people per sq km)

10–14
5–9
0–4

3.8   Using the right diagram for the data ▬▬▬▬

Now you know the different diagrams you can use to display discrete data, you need to be able to choose the best one to use for your data.

This table summarizes the advantages and disadvantages of each diagram:

Diagram	Advantages	Disadvantages
Pictogram	Excellent visual for displaying qualitative data.	Time consuming to draw and not as good for quantitative data. Difficult to read fractions of a symbol.
Bar chart	Simple, easy to read and understand. Can be used for both quantitative and qualitative data.	Only simple information can be shown.
Multiple bar chart	You can compare data quite easily.	You can only show a few categories or it becomes too complex. Many people find them confusing to read.
Composite bar chart	You can compare totals easily.	It is harder to compare categories.
Pie chart	Excellent visual for displaying data with only a few categories.	Hard to draw and the original data is difficult to read from the chart.
Comparative pie chart	Excellent visual way to compare two sets of data.	It only shows a few categories. It is difficult to make numerical comparisons.

Exercise 3H _____

In this exercise you should state which diagram you think is the best one to display the data and give a reason why. You do not need to draw the diagrams.

1. You are collecting data about the populations of towns in England and in Scotland.
    (a) What diagrams could you use to display the information?
    (b) Which one(s) would you expect to be most appropriate? Why?

2. The colour of cars in a car park was recorded.
   (a) Which diagram would you select to represent the data?
   (b) Explain your choice.

3. The number of students absent from school each day was recorded.
   The number ranged from 0 to 15.
   (a) How would you represent the data?
   (b) Explain why you selected some diagrams and rejected others.

4. A student was collecting data on the cost of houses around where
   he lived.
   (a) How could you represent the data?
   (b) Which diagram is most appropriate?
       Why?

5. Katherine carried out a survey comparing the length of time that
   male and female staff had been at her school.

   (a) How could she represent the data?
   (b) Which diagram would you find the most instructive?

3.9   Displaying continuous data

You can use many diagrams to display continuous data. These are the
main ones you will need to know for GCSE Statistics:

✦ Population pyramids
✦ Stem and leaf diagrams
✦ Frequency diagrams and frequency polygons
✦ Histograms
✦ Cumulative frequency polygons

Remember that continuous data is usually collected in groups, so most
of these diagrams use groups to display the data.

3.10   Population pyramids

You can compare two sets of data using a population pyramid.

A population pyramid is useful for comparing two sets of continuous data.
It looks like a bar chart with the bars back to back.

Example

Use a population pyramid to illustrate this data showing the distribution of supporters at a football match. (All data has been rounded to the nearest 10.)

Age (years)	Men	Women
0–9	380	270
10–19	890	620
20–29	1250	970
30–39	1140	510
40–49	830	210
50–59	740	80
60–	680	50

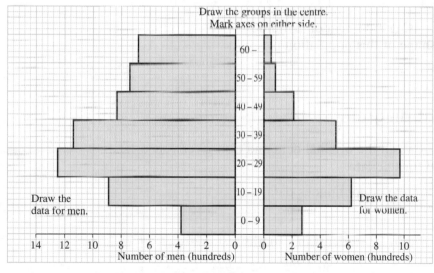

Note that the final group in the table doesn't have an endpoint.
Here we have chosen 69 as the endpoint so the rectangle is the same width as for the other groups.

Now you can compare the distributions at a glance.

Exercise 3I

1. The age distributions for two suburbs of a city, St Marks and Old Town, are as follows:

	0–14	15–29	30–44	45–59	60–74	75 +
St Marks	1000	1660	950	1120	1690	680
Old Town	1110	1570	1380	1710	1250	550

Represent the data using a population pyramid.

2. The percentage population in Brazil by age and gender is given in the following table.

	0–9	10–19	20–29	30–39	40–49	50–59	60–69	70–79	80 +
Males	13.8	12.0	8.2	6.0	4.3	3.9	2.8	1.1	0.3
Females	12.4	11.2	8.5	5.8	4.1	2.8	1.8	0.8	0.2

(a) Represent the data using a population pyramid.
(b) Does a pyramid of this shape show an increase in population or a decrease in population?
(c) What is the shape of a population pyramid if the population is stable?

3. The percentage of Turkish immigrants by age is given in the table below:

	0–9	10–19	20–29	30–39	40–49	50–59	60–69
Males	11.4	7.8	22.3	15.2	5.8	2.8	0.9
Females	11.2	5.3	9.7	4.3	1.2	1.3	0.8

(a) Represent the data using a population pyramid.
(b) Comment on the shape of this population pyramid.

3.11   Stem and leaf diagrams

A stem and leaf diagram shows all the original data and also gives you the overall picture or trend for the data.

You can use it to display discrete or continuous data.

Example

These are the results obtained by 23 students in a maths test.

> 54   75   63   80   63   77   78   86   72   62   94   84
> 87   66   93   56   80   86   51   78   68   73   82

Show this data using a stem and leaf diagram.

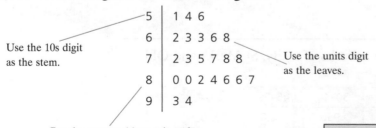

Use the 10s digit as the stem.

Use the units digit as the leaves.

```
5 | 1 4 6
6 | 2 3 3 6 8
7 | 2 3 5 7 8 8
8 | 0 0 2 4 6 6 7
9 | 3 4
```

Put the stem and leaves in order so you can see the pattern clearly.

Key: 5 | 1 means 51.

5 | 1 could mean 5.1 if you are using decimals, so it is important that you remember to give a key.

You can use groups other than 10s for the stem so long as all the groups are the same size. For example, groups of 5 would look like this:

```
5 | 1 4
5 | 6
6 | 2 3 3
6 | 6 8
7 | 2 3
7 | 5 7 8 8
8 | 0 0 2 4
8 | 6 6 7
9 | 3 4
```

Key: 5|1 means 51.

> The groups must be the same width for you to be able to get a sense of the shape of the data.

Exercise 3J

1. The mass in grams of 15 small cakes is as follows:

37 28 36 40 37
29 42 34 31 29
38 32 37 40 30

Represent the data using a stem and leaf diagram.

2. The number of points gained by premier league teams at the start of May 2001 were

38 62 33 46 24
53 47 66 41 62
38 80 46 45 51
54 39 52 62 34

Represent the data using a stem and leaf diagram.

3. The lengths (to the nearest mm) of 27 worms are

58 49 52 57 42 56
55 47 48 43 48 52
59 46 41 59 54 48
53 46 44 55 59 60
46 52 51

Represent the data using a stem and leaf diagram.

3.12   Frequency diagrams and frequency polygons

A **frequency diagram** is similar to a bar chart but:

✦   The bars are drawn with no gaps (as they represent continuous data).

✦   The axis has a continuous scale (rather than distinct categories).

> A frequency diagram uses bars to display grouped data on a continuous scale.

> To draw a frequency polygon you join the midpoints of the bars of a frequency diagram with straight lines.

● Example

This data shows the ages of people attending a Madonna concert. The numbers are rounded to the nearest 100.

Age	0–	10–	20–	30–	40–	50–	60–70
Frequency	2200	18 700	17 500	13 600	9100	6900	1300

Draw a frequency diagram and a frequency polygon to illustrate this data.

To draw a frequency diagram, first work out the best scale to use:

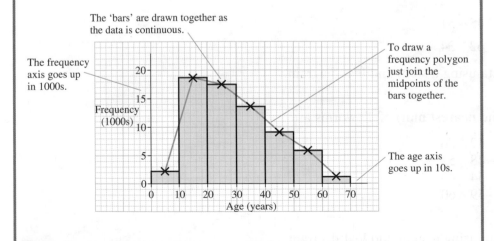

The 'bars' are drawn together as the data is continuous.

The frequency axis goes up in 1000s.

To draw a frequency polygon just join the midpoints of the bars together.

The age axis goes up in 10s.

Exercise 3K _____

1. The temperatures recorded one summer were as follows:

Temperature °F	68	69	70	71	72	73	74	75	76	77	78
No. of days	3	5	12	14	13	8	5	9	6	3	2

Draw (a) a frequency diagram,
      (b) a frequency polygon.

2. The length of pupils forearms was recorded as:

Length of forearm	24	25	26	27	28	29	30	31	32	33
No. of pupils	8	9	21	24	20	27	16	22	12	9

Draw (a) a frequency diagram,
      (b) a frequency polygon.

3. The year of manufacture of 100 cars in a car park is:

Year	1994	1995	1996	1997	1998	1999	2000
No. of cars	42	48	57	79	74	85	52

Draw (a) a frequency diagram,
      (b) a frequency polygon.

4. The table gives the time taken for students to travel to school.

Time (minutes)	$0 \leqslant t < 10$	$10 \leqslant t < 20$	$20 \leqslant t < 30$	$30 \leqslant t < 40$
Frequency	12	36	8	5

Draw (a) a frequency diagram,
      (b) a frequency polygon.

5. The actual time between tube trains at rush hour was recorded to the nearest minute:

Time (min)	7	8	9	10	11	12
No of tube trains	13	18	35	12	19	3

Draw (a) a frequency diagram,
      (b) a frequency polygon.

3.13   Cumulative frequency polygons

A cumulative frequency polygon shows the trend of growth of
continuous data. It is useful for estimating how much more or less there
is than a certain amount.

To draw a cumulative frequency polygon you plot the cumulative
frequency against the **upper boundary** of each class. You then join the
points with straight diagonal lines.

> A **cumulative frequency
> curve** joins the points
> with a smooth curve
> rather than straight lines.

Example

Shereen is collecting data on rats for the local council. She has collected this data on the
length of rats (measured to the nearest cm from nose to tail) found in the area:

Length of rat (cm)	−14	15–18	19–22	23–26	27–30	31–34	35–38	39–
Frequency	3	12	14	21	15	12	7	3

(a) Draw a cumulative frequency polygon for the data.
(b) Use your polygon to estimate the percentage of rats over
     25 cm long.

> Note that in the last category
> you have to choose an
> appropriate boundary.

(a) First make a table showing the upper class boundaries against the cumulative frequencies.
     The upper class boundary of the first class is between 14 and 15, so it is 14.5.

Length of rat (cm)	−14	15–18	19–22	23–26	27–30	31–34	35–38	39–
Upper class boundary (cm)	14.5	18.5	22.5	26.5	30.5	34.5	38.5	42.5
Frequency	3	12	14	21	15	12	7	3
Cumulative frequency	3	3 + 12 = 15	15 + 14 = 29	29 + 21 = 50	50 + 15 = 65	65 + 12 = 77	77 + 7 = 84	84 + 3 = 87

(b) Plot the points:

(14.5, 3)   (18.5, 15)   (22.5, 29)   (26.5, 50)   (30.5, 65)   (34.5, 77)   (38.5, 84)   (42.5, 87)

Join them up with straight lines:

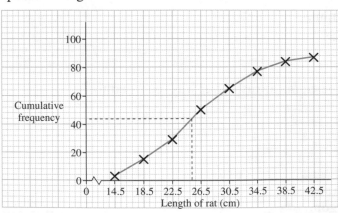

> (b) The number of rats under
> 25 cm long is approximately 44
> from the graph. The total number
> of rats in the survey is 87, so the
> number of rats over 25 cm is
> approximately 43. The percentage
> is $\frac{43}{87} \times 100\% = 49.4\%$.

Exercise 3L _____

1. The table gives the time taken for students to travel to school.

Time (minutes)	$0 \leqslant t < 10$	$10 \leqslant t < 20$	$20 \leqslant t < 30$	$30 \leqslant t < 40$
Frequency	12	36	8	5

Draw a cumulative frequency graph for the data.

2. The following table gives the time taken to solve a simple jigsaw puzzle.

Time (seconds)	$0 \leqslant t < 30$	$30 \leqslant t < 45$	$45 \leqslant t < 60$	$60 \leqslant t < 75$	$75 \leqslant t < 90$	$90 \leqslant t < 110$	$110 \leqslant t < 180$
Frequency	4	12	25	32	27	22	12

Draw a cumulative frequency graph for the data.

3. The table gives the age distribution (in complete years) of a village.

Age	0–9	10–19	20–24	25–29	30–39	40–59	60–79	80 and over
Frequency	38	47	25	19	21	24	18	5

Draw a cumulative frequency graph for the data.

4. The cumulative frequency graph shows the length of time, in seconds, it takes a group of 80 students to solve a particular problem.

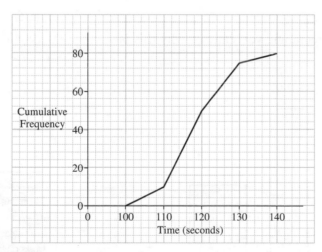

Time (seconds)	$100 \leqslant t < 110$			
Frequency	10			

Copy the table and use the graph to fill in the missing entries.

3.14   Histograms

In a frequency diagram the heights of the bars represent the frequencies of the groups.

A histogram is similar to a frequency diagram but instead it is the **areas** of the bars that represent the frequencies of the groups.

A histogram may have equal or unequal intervals.

Equal intervals

> A histogram with equal intervals is the same as a frequency diagram.
> The width of each bar is exactly the same and so only the heights vary.

Unequal intervals

> In a histogram with unequal intervals, the area of each bar is proportional to the frequency of each class.

> The height of each bar is called the frequency density.

To find the frequency density, you divide the frequencies by the class widths.

If you think about the area of a rectangle you can see why this is important:

These rectangles all stand for 24 people:

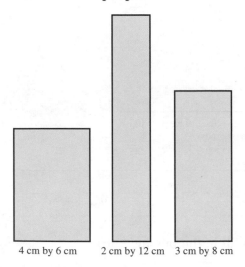

4 cm by 6 cm   2 cm by 12 cm   3 cm by 8 cm

> They all have an area of 24 cm^2 but they are all different sizes.
> In a histogram, any of these blocks could represent 24 people.

Example

This data shows the heights of flowers in a flowerbed.

Height (cm)	0–8	8–12	12–16	16–18	18–20	20–24	24–28	28–36
Frequency	2	9	12	8	7	6	5	2

Draw a histogram to show the data.

First draw a table to show the widths and frequency densities of the classes (and hence the required heights).

Height of flower (cm)	0–8	8–12	12–16	16–18	18–20	20–24	24–28	28–36
Class width	8	4	4	2	2	4	4	8
Frequency	2	9	12	8	7	6	5	2
Frequency density Height of bar — frequency ÷ class width	$2 \div 8$ $= 0.25$	$9 \div 4$ $= 2.25$	$12 \div 4$ $= 3$	$8 \div 2$ $= 4$	$7 \div 2$ $= 3.5$	$6 \div 4$ $= 1.5$	$5 \div 4$ $= 1.25$	$2 \div 8$ $= 0.25$

Now plot the height of the flower against the **frequency density**.

Note that this axis is no longer the frequency but is now called the **frequency density**.

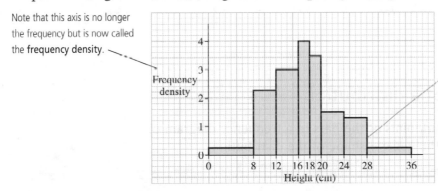

Check that the areas are equal to the frequencies. For example for this bar: $4 \times 1.25 = 5$, which is the frequency for that class.

Exercise 3M

1. The following table gives the time taken to solve a simple jigsaw puzzle.

Time (seconds)	$0 \leqslant t < 30$	$30 \leqslant t < 45$	$45 \leqslant t < 60$	$60 \leqslant t < 75$	$75 \leqslant t < 90$	$90 \leqslant t < 110$	$110 \leqslant t < 180$
Frequency	4	12	25	32	27	22	12

Draw a histogram to represent this data.

2. Use the data from Exercise 4K, question 1 on page 112.
Draw a histogram of the data.

3. The table gives the age (in complete years) distribution of a village.

Age	0–9	10–19	20–24	25–29	30–39	40–59	60–79	80 and over
Frequency	38	47	25	19	21	24	18	5

Make a sensible assumption about the last interval and construct a histogram to represent the data.

4. The table gives the heights of 180 students.

Height (nearest cm)	140–144	145–149	150–152	153–154	155–164
Frequency	36	54	43	22	25

Draw a histogram of the data.

5. The age distribution of people living in The Vale is given in the table:

Age last birthday	0–9	10–14	15–19	20–29	30–39	40–59	60–79
Frequency	32	25	28	22	35	52	41

Draw a histogram of the data.

3.15 The shape of a distribution

If you join together the midpoints of the bars in a frequency diagram or a histogram, you can see the general shape of the distribution.

The narrower the bars, the clearer the shape.

There are some shapes that occur frequently in statistics and you need to be able to recognize and name them.

Symmetrical distribution
This describes a distribution which is more or less symmetrical.

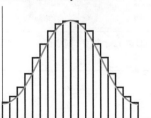

Real-life examples include …
• the lengths of leaves on a particular tree
• the heights of a random sample of people.

Positive skew
This describes a distribution with most of the data at the lower values.

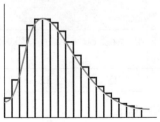

Real-life examples include …
• the age at which a sample of people first learned to read
• the heights of jockeys in a horse race.

Negative skew
This describes a distribution with most of the data at the higher values.

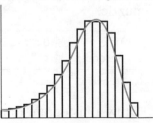

Real-life examples include …
• the age at which a sample of people had false teeth fitted
• the heights of players in a basketball team.

Exercise 3N _____

1. For question 1 in Exercise 3M describe the shape of the distribution.

2. For question 2 in Exercise 3M describe the shape of the distribution.

3. For question 3 in Exercise 3M describe the shape of the distribution.

4. For question 4 in Exercise 3M describe the shape of the distribution.

5. For question 5 in Exercise 3M describe the shape of the distribution.

Summary

You should now be able to	Check out 3
1 Draw a variety of simple diagrams.	1 The number of meals sold in a restaurant is given. Draw a pie chart and a bar chart to represent the data. Meal / Frequency: Fish and chips 58, Roast dinner 25, Pizza 37
2 Decide which is the most appropriate diagram to use.	2 (a) What type of chart would you use to show absolute values? (b) What type of chart would you use to illustrate proportions?
3 Draw cumulative frequency graphs.	3 Draw a cumulative frequency graph to illustrate the following data for lengths of palm leaves. Length (cm) / Frequency: $20 \leqslant L < 40$ — 24, $40 \leqslant L < 50$ — 28, $50 \leqslant L < 60$ — 30, $60 \leqslant L < 80$ — 36, $80 \leqslant L < 110$ — 24
4 Draw histograms.	4 Draw a histogram to represent the data in question 3.

Revision Exercise 3

1. The diagram shows the percentage of teachers by age in some European countries.

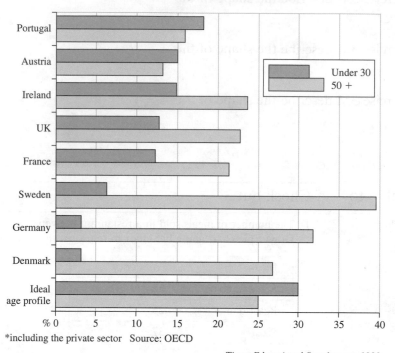

HOW OLD ARE EUROPE'S TEACHERS?

Percentage of teachers by age in primary and secondary education* 1994–95

*including the private sector   Source: OECD

Times Educational Supplement, 1999

(a) What percentage of teachers from Austria are under 30 years of age?

(b) What percentage of teachers from Sweden are over 50 years of age?

(c) What percentage of teachers from Germany are between 30 and 50 years of age?

(d) Which of the following countries, Portugal, Austria, Ireland and UK has the greatest percentage of teachers between 30 and 50 years of age?   [NEAB]

2. The diagram below shows how a large company spent its advertising budget in 1996.

Advertising Budget for 1996

Newspapers 55%	Television	Direct mail 10%	Others 8%

NOT DRAWN TO SCALE

0   20   40   60   80   100%

(a) What percentage of the advertising budget was spent on television advertising?

(b) The total advertising budget was £400 000. How much was spent on newspaper advertising?   [NEAB]

3. The pie chart below shows the colours of the flowers produced by the tulip bulbs planted in a garden.

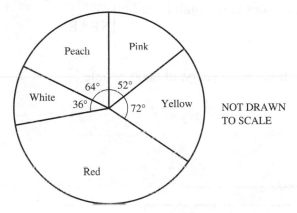

NOT DRAWN
TO SCALE

(a) There are 9 bulbs producing white flowers. How many tulip bulbs are there in the garden altogether?

(b) How many of the tulip bulbs produce red flowers?   [NEAB]

4. The diagrams below show the percentage of the whole population of the United Kingdom by age and sex for the years 1901 and 1991.

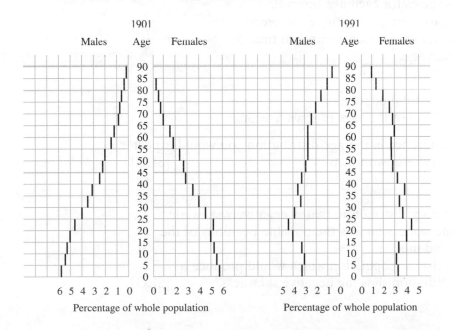

Source: Adapted from *Annual Abstract of Statistics 1992*.

(a) (i) In 1901, what percentage of the population was male and
under 15 years?
(ii) In 1991, what percentage of the population was female
and in the age group 20 to 40 years?
(b) (i) Comment on the differences in the age structure of the
population between 1901 and 1991.
(ii) Give a reason for these differences.
(c) Compare the percentages of males and females aged 60–90
years in 1991.   [NEAB]

5. The population pyramid shows the ages, in years, of 6000 people in
town X.

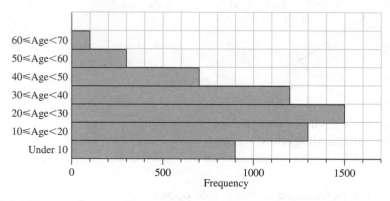

(a) How many people are between 20 and 40 years of age?
(b) List the frequencies for each age interval.
(c) Draw a cumulative frequency curve for age.
(d) Use your cumulative frequency curve to find
(i)   the median age,
(ii) the interquartile range.

Town **Y** has a similar total population.
The interquartile range of the ages of town **Y** is 30.

(e) Explain **one** difference in the age distribution you would expect
to find between these two towns.   [SEG]

6. (a) Sketch a frequency distribution that is negatively skewed.
(b) Indicate on the horizontal axis the possible positions of the
mean, mode and median.
(c) Describe a statistical population which would produce a
frequency distribution showing negative skewness.   [SEG]

7. Information about the schools provided by the Ministry of Defence is shown in the diagram.

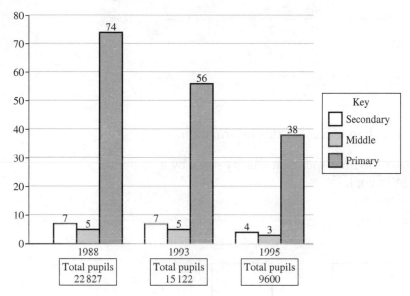

(a) How many schools did the Ministry of Defence provide in 1988?

(b) Calculate, to the nearest whole number, the average number of pupils per school for the year 1995.

The rate of decline in the number of schools over the period 1988 to 1993 was 3.6 schools per year.

(c) Calculate the rate of decline in the number of schools over the period 1993 to 1995.

(d) Explain how the pictorial representation could easily lead to a misinterpretation of the actual information given.

8. Each month 240 people do some work for charity. The pie chart represents the time, t, spent by these people on their charity work.

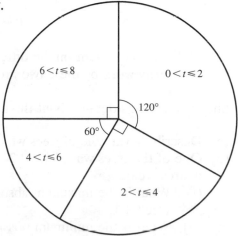

(a) (i)   Copy and complete the frequency table for these data.

	Time, t (hours)	Frequency
	$0 < t \leqslant 2$	
	$2 < t \leqslant 4$	
	$4 < t \leqslant 6$	
	$6 < t \leqslant 8$	

(ii)   Draw a bar graph to represent this information.

The graph below shows the time spent on charity work by a different sample of 240 people.

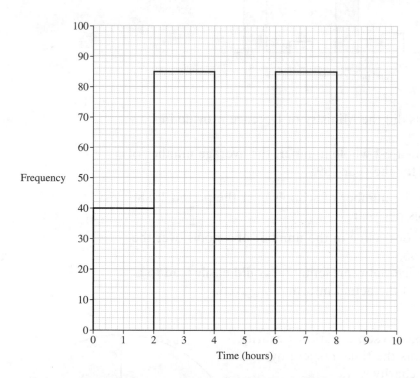

(b) Make **one** comment on the difference between the time spent on charity work by these two groups of people.   [SEG]

9. Sheridan collects details about the trees in a wood.

(a) Describe a variable of trees which is quantitative.
(b) One of the trees was planted 24 years ago, correct to the nearest year.
    (i)   What is the maximum absolute error in the age of this tree?
    (ii) What is the minimum possible age of the tree?

(c) The pie chart below shows the proportion of the different types of tree recorded.

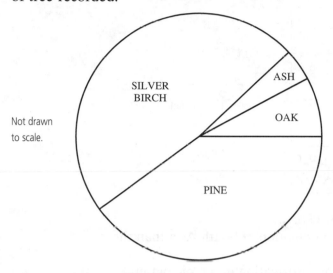

Not drawn to scale.

There are 9 Oak trees and these are represented by a sector with an angle of 27°.
(i) The Pine trees are represented by an angle of 144°. How many Pine trees are there?
(ii) There are 58 Silver Birch trees. Calculate the angle of the sector representing Silver Birch trees on the pie chart.
(d) Pine trees are evergreen. Ash, Silver Birch and Oak are deciduous. What is the ratio of evergreen trees to deciduous trees?   [SEG]

10. The table shows the amount of pocket money received, each week, by 720 young children. Every child received an exact number of pounds.

(a) The information is to be shown on a pie chart. Copy and complete the table.

Pocket money per week	Number of young children	Angle on pie chart
£1	120	60°
£2	180	
£3	160	
£4	200	
£5 or more	60	

(b) Draw a pie chart to illustrate these data.   [NEAB]

11. The pictogram shows populations of some regions in 1970.

U.S.A.	
Europe	
South America	
South Asia	

Scale = 80 million people

(a) What was the population of U.S.A.?
(b) How much greater was the population of South Asia than the population of Europe?
(c) Draw a pie chart to show the information in the pictogram.
                                                                    [NEAB]

12. The following table shows the annual salaries of the 100 employees of a small manufacturing company.

Salary (in £000s) x	Frequency
$5 \leqslant x < 10$	10
$10 \leqslant x < 12$	10
$12 \leqslant x < 14$	22
$14 \leqslant x < 15$	21
$15 \leqslant x < 17$	18
$17 \leqslant x < 20$	12
$20 \leqslant x < 27$	7

(a) Draw a histogram to represent these data.
(b) Use your histogram to identify the modal class for this distribution.
(c) Calculate the probability that, of two randomly selected employees, both earn annual salaries in the range £10 000–£13 999.
                                                                    [NEAB]

You can find out about Probability in Unit 7.

13. The table below shows the amount of fat consumed per day, in grams, on average by someone from the United Kingdom and someone from Italy.

	United Kingdom	Italy
Saturated Fat	35	15
Mono Unsaturated Fat	24	22
Poly Unsaturated Fat	22	56

(a) Draw a composite bar chart to illustrate this information.
(b) Describe briefly the differences in fat consumption between the United Kingdom and Italy.   [NEAB]

14. (a) The money spent on chilled foods in Britain, in 1995, is shown on the diagram.

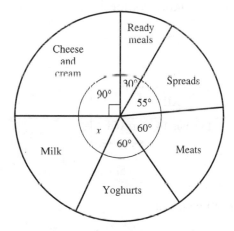

(i)   Calculate the size of angle x.
(ii) Sales of cheese and cream totalled £270 million.
       How much was spent on ready meals?

(b) The sales of some types of cooked meats, for the same year,
are shown in the table below.

Cooked meats	Sales in millions of pounds	Angle on pie chart
Turkey	48	
Chicken	22	
Beef	8	
Luncheon meat	12	
Total	90	

The information is to be shown on a pie chart.
Find the missing angles. Copy and complete the table.
Draw a pie chart to show this information.   [NEAB]

15. The stem and leaf diagram below shows the ages, in years, of
26 people who wished to enter a 10-mile walking competition.

stem	leaf
0	3 4
1	4 4 6 9
2	1 3 6 6 6 8
3	3 5 5 7 9
4	0 2 3 3 7 7
5	1 2 4

KEY
1 4 means 14 years old

(a) How many people were less than 20 years old?
(b) Write down the modal age.
(c) Exactly half the people entering the walk are more than a
certain age. What is that age?
(d) Shaun says that two people will not be allowed to enter.
Using the information in the stem and leaf diagram,
suggest a reason for this.   [NEAB]

16. The charts show the amount, in pounds, spent in the week before Christmas by shoppers at a food store in 1996 and 1999.

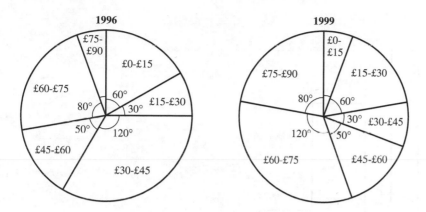

1996   1999

(a) In **1999**, which range of amounts was most likely to contain the modal amount spent?

(b) Which range of amounts was spent by the same fraction of shoppers in both 1996 and 1999?

(c) State **one** change in the spending pattern shown by the charts between these years.

In 1996 there were 2700 shoppers.

(d) Copy and complete the frequency table for 1996.

Amount spent (£)	Angle (degrees)	Frequency	
0 and less than 15	60	450	
15 and less than 30	30		
30 and less than 45	120		
45 and less than 60	50		
60 and less than 75	80		
75 and less than 90			

(e) Draw a histogram to represent the amounts spent in 1996.

[SEG]

17. The histogram shows the journey time taken by 91 factory workers to get to work.

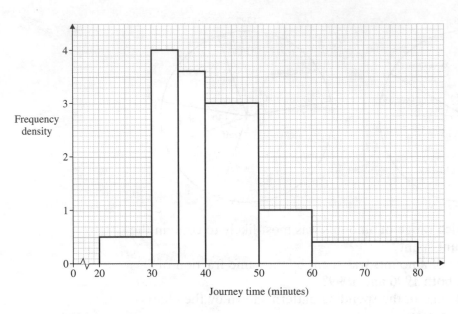

(a) Describe the skewness of the distribution.
(b) Calculate the number of workers whose journey time was in the interval 30–35 minutes.
(c) Calculate an estimate of the median journey time.   [SEG]

18. The table gives the mean monthly rainfall, in mm, for Bali for the periods 1951–80 and for 1989–91.

Years	Jan & Feb	Mar & Apr	May & Jun	Jul & Aug	Sep & Oct	Nov & Dec
1951–80	110	81	75	90	115	127
1989–91	167	107	54	52	103	150

(a) Draw comparative bar charts of the two sets of data.
(b) (i) Describe the main difference between the two sets of data.
    (ii) Which is the more reliable set of data? Give a reason for your answer.   [SEG]

4 Measures of location and spread

Dogs come in all shapes and sizes, but some look more 'typical' than others.

This unit will show you how to

+ Calculate averages for simple data
+ Use graphs to obtain quartiles
+ Calculate mean and standard deviation for grouped data
+ Use standardized scores

Before you start

You need to know how to	Check in 4
1 Use decimals	**1** (a) 0.5×0.7 (b) $0.03 + 0.4$ (c) $0.75 - 0.643$

2 Find upper and lower class bounds	**2** (a) A measurement is recorded to the nearest centimetre in the interval $50 < L < 59$.
	      Write down the lower class bound and the upper class bound.
	     (b) A person's age is recorded in **complete** years in the interval 50–59.
	      Write down the lower class bound and the upper class bound of the age.

3 Use formulae	**3** Use the formula $x = \frac{(a+b+c)}{3}$ to find the value of x when $a = 4$, $b = 8$ and $c = 9$.

4.1   Mean, median and mode

> An **average** is a single value that can be used to describe a set of data.

When you collect information you can use averages to summarize your information.

Rachel and Dave own a clothes shop and they keep a record of their sales. They want to know their average daily takings.

Here is a table showing their takings over a particular fortnight.

All figures are given to the nearest pound.

First week:

Mon	Tue	Wed	Thu	Fri	Sat
95	129	130	106	130	594

Second week:

Mon	Tue	Wed	Thu	Fri	Sat
110	125	132	118	130	468

> Averages are sometimes called **central values**, or measures of location.

There are three types of average: mode, median and mean.

The mode

> The mode is the most frequently occurring value.

● **Example**

In the tables of the shop's takings, 130 appears most often.

The mode is £130.

> If the amounts were all different, there would not be a mode. If two values occur most frequently the data is bimodal.

The median

> The median is the middle value when the data is arranged in numerical order.
>
> For a list of n items, the median number is the $\frac{n+1}{2}$ th item.

Example

When written in order, the numbers in the tables are:

95, 106, 110, 118, 125, 129, 130, 130, 130, 132, 468, 594

There are 12 items, so $n = 12$.

The median is the $\left(\dfrac{12+1}{2}\right)$ th number, that is the $6\frac{1}{2}$ th number, so take the value $\frac{1}{2}$ way between the 6th and 7th values.

The median is £129.50.

The arithmetic mean

You find the arithmetic mean by:
+ Adding up all the data.
+ Dividing the total by the number of items.

Note:
The arithmetic mean is often just called the mean.

Example

To find the mean of the clothes shop's data:
+ Add all the data:
  $$95 + 106 + 110 + \ldots + 594 = 2267$$

+ Divide the total by the number of items:
  $$2267 \div 12 = 188.916\ldots$$
The mean is £189 to the nearest pound.

All three averages are different. Which should Rachel and Dave choose? You will find out in section 4.6.

Exercise 4A

1. Find (i) the mode, (ii) the median, and (iii) the mean of the following data sets:

    (a) 6, 7, 3, 8, 4, 8, 3, 2, 3
    (b) 12, 15, 18, 15, 14, 17
    (c) 22, 23, 18, 20, 25, 23, 27

2. The marks obtained, in a spelling test, by two groups of people are given in the following table:

Group 1	46	50	46	52	46	55	58	54	60
Group 2	39	72	39	68	48	74	39	52	75

    Find (a) the mode, (b) the median and (c) the mean of both groups of data. (d) Compare your findings and comment on who was better at spelling.

3. The weekly pocket money of a group of students is recorded below:

£8, £3, £5, £4, £7.50, £8, £5.50, £5, £10.

(a) Is there no mode or is the data bimodal?

(b) Put the amounts in order and find the median.

(c) Calculate the mean weekly pocket money.

Another student joins the group. She gets £8 per week pocket money.

(d) Find the mean, median and mode for all 10 students.

(e) Has the average pocket money increased or decreased? Explain why.

4. The mean height of a group of eight students is 165 cm.

(a) What is the total height of all eight students?

A ninth student joins the group. He is 168 cm tall.

(b) What is the mean height of all nine students?

4.2   Averages from discrete frequency tables

Parveen is doing a project on the number of people in a household. She has designed an observation sheet and taken it around her class. She wants to find the average size of a household.

Example

Find the mean, median and mode of the data from Parveen's table:

Number of people in household	Tally	Frequency	Running total
2	III	3	3
3	JHT I	6	9
4	JHT III	8	17
5	JHT I	6	23
6	III	3	26
7	II	2	28
8		0	28
9	II	2	30

Running total is usually known as **cumulative frequency**, which is explained on page 61.

To find the **mode**, look at the highest frequency. The highest frequency is 8. **The mode is 4.**

To find the median, find the $\frac{1}{2}(n+1)$th value. $n = 30$, so the median is the $15\frac{1}{2}$th value. The 10th to 17th values are 4. So the 15th and 16th values are both 4. **The median is 4.**

So the mode and the median are both 4 people.

To find the mean you can rewrite the frequency table without the tally column. You will need an extra column to give you the total number of people in households.

Number of people (x)	Frequency (f)	fx
2	3	$3 \times 2 = 6$
3	6	$6 \times 3 = 18$
4	8	$8 \times 4 = 32$
5	6	$6 \times 5 = 30$
6	3	$3 \times 6 = 18$
7	2	$2 \times 7 = 14$
8	0	$0 \times 8 = 0$
9	2	$2 \times 9 = 18$
Totals	30	136

There are 3 students with 2 people in their household. This makes 6 people in total.

This is the total number of people in all households in the survey.

This is the total number of students.

The mean number of people per household is given by $\dfrac{\Sigma fx}{\Sigma f}$.

$136 \div 30 = 4.53$ (to 2 d.p.).

The symbol Σ means 'sum of'. See page 94.

Exercise 4B _____

1. Find the mode, median and mean in each of the following cases.
   (i)   7, 4, 9, 12, 7, 4, 11, 8
   (ii) 5.9, 5.7, 5.5, 5.5, 6.0, 5.5, 5.9, 5.9

2. The mean of ten numbers is 6.7. What is the total of the 10 numbers?

3. The mean height of 8 boys is 1.54 m.
   The mean height of 6 of the boys is 1.48 m.
   What is the mean height of the other two boys?

4. The mean weight of 12 parcels is 2.58 kg.
   The mean weight of a further 8 parcels is 3.13 kg.
   What is the mean weight of all 20 parcels?

5. The table gives the length of words in a crossword puzzle.
   Obtain values for the mode, median and mean length of word.

Length of word	Frequency
4	5
5	8
6	9
7	4
8	5
9	8
10	4
11	2

6. The number of matches in a box was counted for a sample of
40 boxes.
The results are summarized in the table.
Obtain values for the mode, median and mean number of matches
in a box.

Number of matches	Frequency
31	1
32	5
33	6
34	7
35	8
36	7
37	6

7. Parveen found an error in her original data (see the example at the
start of the section). One of the students forgot to include himself
when giving his answer. Instead of three people in his household
there are in fact four.
Recalculate the mean, median and mode for her data.

4.3 Calculating the mean by formula

In algebra you use letters to represent numbers. You can use algebra
in statistics as well.

Tyrone is trying to explain to Lee how to find the mean of a list of
numbers. Lee is rather unsure of Tyrone's rule. Even though it is
correct, it is confusing in words.

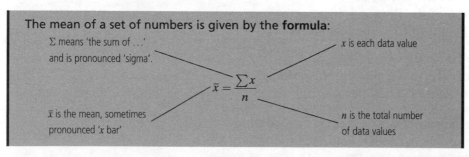

Example

The total of a set of 12 numbers is 32.

What is the mean?

Using the formula, $\Sigma x = 32$ and $n = 12$.

$$\bar{x} = 32 \div 12$$
$$= 2\tfrac{2}{3}.$$

When data is given in a frequency table the formula is slightly different.

> The mean of a **frequency distribution** is given by the formula:
> $$\bar{x} = \frac{\sum fx}{\sum f}$$

Example

Sally recorded the following data on the number of peas contained in 50 pods.

Number of peas	5	6	7	8	9	10
Number of pods	8	11	13	7	6	5

Find the mean number of peas per pod.

First rewrite the table with x as the number of peas and f as the number of pods. Add another column for fx.

x	f	fx
5	8	40
6	11	66
7	13	91
8	7	56
9	6	54
10	5	50
Total	50	357

$$\sum fx = 357 \qquad \sum f = 50$$

$$\bar{x} = \frac{357}{50} = 7.14$$

The mean number of peas per pod is 7.14.

Exercise 4C

1. The number of pupils in a class in a particular school is given in the following table:

Number of pupils	25	26	27	28	29	30	31	32
Number of classes	1	2	3	5	2	1	0	1

Find the mean number of pupils per class.

2. The number of GCSEs gained by a random sample of 100 students was recorded.

Number of GCSEs	0	1	2	3	4	5	6	7	8	9	10
Number of students	4	6	7	9	13	15	12	10	12	7	5

Calculate the mean number of GCSEs per student.

3. The mean number of goals per game scored by a certain football team during a season is 1.8.
   They have scored 72 goals in total. How many matches have they played?

4. The populations of six villages in a particular parish are given as:

   526, 1048, 381, 620, 305 and 212.

   (a) Find the mean population.
   (b) How would you describe the *spread* of population?

4.4   Averages from continuous data

So far in this unit you have looked at discrete data. You can also work out averages for continuous data. You use the midpoint of the class interval as your *x* value as it is in the middle of the data and so more likely to be typical.

> You can find out more about discrete and continuous data on page 2.

Mean of a grouped frequency table

● **Example**

Charmaine is doing a project on speed of writing. She wants to find the average time taken for the students in her class to write a chosen passage of 400 words. She records their times in a table.

Time taken (minutes)	0–8	8–10	10–12	12–16	16–20
Number of students (f)	2	4	12	5	2

Use this data to estimate the mean length of time to write a passage.

Rewrite the table:

Time to complete (to nearest minute)	Mid-value x	f	fx
0–8	4	2	8
8–10	9	4	36
10–12	11	12	132
12–16	14	5	70
16–20	18	2	36
Totals		25	282

$$\sum f = 25 \qquad \sum fx = 282$$

$$\bar{x} = \frac{\sum fx}{\sum f}$$

$$= 282 \div 25$$

$$= 11.28$$

The mean time is 11.3 minutes (to 1 decimal place).

> Note that this is only an **estimate** of the mean. It cannot be calculated accurately as you have not been given the raw data.

Mode and median of a grouped frequency distribution

With grouped data you cannot give exact values of the mean, median or mode. You can only estimate them.

> The **modal class** is the interval in which the mode lies. It is the class with the highest frequency.
>
> When the total frequency is large, you can approximate the median as the halfway value, or $\frac{n}{2}$th value.

> To find the average of a set of grouped data, you often use the midpoint of the data:
>
> ✦ For the mean, you use the midpoint of each group.
>
> ✦ For the median, you find the midway value.
>
> This is because the middle value is usually more typical than the values at the extremes.

Example

This data shows the lengths of a sample of 40 hamsters.

Length of hamster (to nearest cm)	4–	6–	8–	10–	12–14
Frequency	3	15	12	8	2

Find the modal class and estimate the median by drawing a cumulative frequency polygon.

The highest frequency is 15, and this corresponds to the 6–8 interval.

The modal class is 6–8 cm.

The cumulative frequencies are 3, 18, 30, 38 and 40.

You plot the cumulative frequencies against the upper class boundaries:

(4, 0), (6, 3), (8, 18), (10, 30), (12, 38) and (14, 40).

Here is the cumulative frequency polygon:

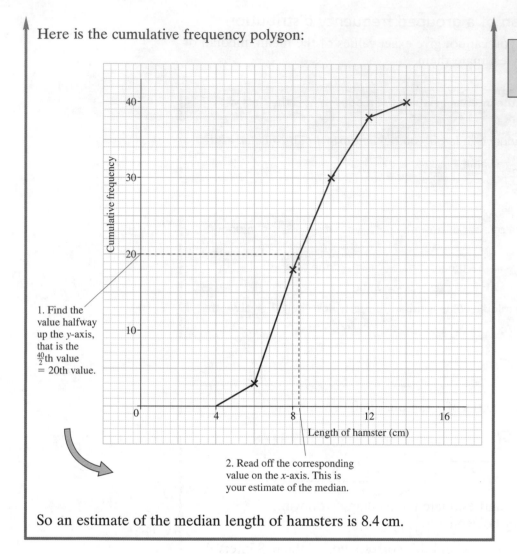

Cumulative frequency polygons are explained on page 72.

1. Find the value halfway up the y-axis, that is the $\frac{40}{2}$th value = 20th value.

2. Read off the corresponding value on the x-axis. This is your estimate of the median.

So an estimate of the median length of hamsters is 8.4 cm.

Exercise 4D

1. Estimate the mean of these grouped continuous data. Give your answer to one decimal place where appropriate.

(a)
Class	1–5	6–10	11–15	16–20
Frequency	2	9	3	1

(b)
Class	10–	20–	30–	40–	50–60
Frequency	8	11	13	9	7

(c)
Class	10–12	12–14	14–16	16–18	18–20
Frequency	1	5	12	3	0

2. The heights of 50 giant redwood trees were measured and recorded to the nearest metre. The results are given in the table:

Height, h, in metres	$100 < h \leqslant 110$	$110 < h \leqslant 120$	$120 < h \leqslant 130$	$130 < h \leqslant 140$	$140 < h \leqslant 150$
Frequency	6	13	17	12	2

(a) Write down the modal class.

(b) Estimate the mean height of this sample of trees. Give your answer to 1 decimal place.

(c) Write down the cumulative frequencies for this data. Draw a cumulative frequency polygon on 2 mm graph paper, clearly labelling your axes.

(d) Use your graph to estimate the median.

You cannot say what the real heights are because only grouped data is given.

(e) (i) Write down the maximum possible height of a redwood from this sample.

(ii) Write down the minimum possible height.

(iii) What is the greatest possible difference between heights for this sample?

(iv) What is the least possible difference?

4.5 Use of scaling to calculate the mean

When data involves large or awkward numbers you can sometimes make the numbers easier by **scaling**. This means you will make fewer errors in your calculations.

p259
p261

Example

Jez and Ricky are planning to go backpacking in Paraguay, and visit the bank to find out the currency exchange rate.

They record the exchange rate over a week and these are their results (in guarani):

5018, 5006, 5007, 5020, 5009, 5012, 5017

'guarani' is the main currency unit in Paraguay.

Find the mean rate using a method of scaling.

Subtract 5000 from each value: 18, 6, 7, 20, 9, 12, 17

Add the scaled values together: $18 + 6 + 7 + 20 + 9 + 12 + 17 = 89$

Divide by the number of values, in this case 7: $89 \div 7 = 12.71$

Add 5000 to the answer: $12.71 + 5000 = 5012.71$

The mean rate is 5012.71 guarani to the pound.

You can find the mean of a set of large numbers by subtracting a common amount from each data value.	Remember to add the amount back on at the end.

You can also divide to make the numbers easier:

Example

Find the mean of these numbers:

304, 308, 316, 312, 320, 300, 324, 320, 316, 308

Subtract 300 from each number: 4, 8, 16, 12, 20, 0, 24, 20, 16, 8

Notice that you can divide all
the numbers by 4:   1, 2, 4, 3, 5, 0, 6, 5, 4, 2

Add the scaled numbers together: $1 + 2 + 4 + 3 + 5 + 0 + 6 + 5 + 4 + 2 = 32$

Divide by the number of values: $32 \div 10 = 3.2$

Remember to include the zero in your calculations!

Now you need to scale back.

Multiply by 4: $3.2 \times 4 = 12.8$

Add on 300: $12.8 + 300 = 312.8$

The mean of the set of numbers is 312.8.

This method is called **scaling** and is useful when dealing with large numbers and with numbers that contain a common factor.

A flow chart of this process would look like this:

data values $\rightarrow$ -300 $\rightarrow$ $\div 4$ $\rightarrow$ scaled values

The reverse flow chart to find the mean would be:

data mean $\leftarrow$ $+300$ $\leftarrow$ $\times 4$ $\leftarrow$ scaled mean

Exercise 4E

1. Find the mean of these sets of data by first subtracting an appropriate amount:

   (a) 1003, 1005, 1008, 1001, 1002
   (b) 508, 502, 509, 513, 518, 506, 514, 512
   (c) 82.5, 82.3, 82.8, 82.4, 82.1, 82.0, 82.1, 82.2

2. The monthly wages of 10 office workers are listed below:

£1009, £1012, £1006, £1003, £1018, £1015, £1000, £1006, £1021, £1009

Find the mean monthly wage by using the method of scaling.

Hint:
After you subtract 1000, look for a **common factor** in your numbers.

3. Use scaling to find the mean of this set of numbers:

414, 442, 484, 449, 428, 407, 435, 491

4.6 Which average to use?

When people use the term 'average', you should try to find out which average they are referring to.

The table gives the main advantages and disadvantages of each average, so you can decide which one to use in a given situation.

Average	Advantages	Disadvantages
Mean	✦ Uses all the data ✦ Most 'accurate' value	✦ Distorted by extreme values ✦ Mean is not always a data value
Median	✦ Unaffected by extremes ✦ Easy to calculate if data is ordered	✦ Not always a data value ✦ Not easy to use for further analysis
Mode	✦ Very easy to find ✦ Can be used with qualitative data ✦ Mode is always a data value	✦ There is not always a mode ✦ Not easy to use for further analysis

● **Example**

The annual salaries of the employees in a small company are listed below:

£25 000, £20 000, £19 000, £17 000, £15 000, £100 000

Find (a) the mean, and (b) the median.

(c) Which average gives the most 'typical' figure, and why?

(d) Why can't you find the mode?

(a) $25\,000 + 20\,000 + 19\,000 + 17\,000 + 15\,000 + 100\,000 = 196\,000$
$196\,000 \div 6 = 32\,666.666\ldots$
The mean is £32 666.67.

> This is an **extreme** value. It distorts the mean.

(b) Arranged in order the amounts are:
15 000, 17 000, 19 000, 20 000, 25 000, 100 000
The median is halfway between the third and fourth values, that is 19 500.
The median salary is £19 500.

(c) Only one employee earns more than the mean and five earn much less.
Three employees earn more than the median and three earn less.
The median is the fairer average to use here.

(d) There is no mode because all of the data values appear just once.

Exercise 4F

1. (a) Find the median of this set of numbers:
   12, 12, 14, 15, 17, 20, 95.

   (b) Why is the median a good choice of average in this case?

2. (a) Find the mean of this set of numbers:
   37, 26, 37, 18, 18, 20, 26, 18, 37, 37, 18.

   (b) Why is the mode a bad choice of average in this case?

3. In March, Rachel and Dave sold five different types of jacket in their clothes shop in the amounts shown.

Jacket	Leather	Suede	Denim	PVC	Cotton
Amount sold	45	17	64	28	52

   (a) Find the modal type of jacket sold.

   (b) Why is the mode appropriate for this data?

4. Jez and Ricky are backpacking in Paraguay and have recorded these daily temperatures for a week.

Day	1	2	3	4	5	6	7
Midday temperature (°C)	32	30	30	28	33	31	30

   (a) Find the mean midday temperature.

   (b) Give a reason why the mean could be appropriate for this data.

Questions 5 to 8 give data on completely different situations. Use the table on page 101 to decide which is the best average to use for each question. Give a reason for your choice, and then calculate the average.

5. Twelve batteries were tested to find out how long they would last. These are the results (in hours to the nearest hour):

    1005, 1019, 1065, 989, 1032, 1008, 1087, 1025, 12, 18, 1013, and 1034.

> This is an example of destructive testing, and it is used commonly in industry.

6. Sixteen teenagers were asked to state their favourite colour. These are the results:

    blue, green, purple, black, purple, pink, white, blue, yellow, pink, black, green, orange, green, black, black.

7. A sample of twenty fish was taken from a pond and their lengths were noted to the nearest centimetre. (Each fish was put back in the pond before the next one was taken out.) The lengths are listed in centimetres below:

    12, 14, 15, 12, 18, 15, 11, 19, 16, 14, 15, 17, 10, 19, 18, 15, 14, 17, 19, 13.

8. The age at which a sample of 15 drivers passed their driving test is recorded below:

    18, 20, 25, 26, 17, 48, 19, 21, 22, 24, 31, 52, 23, 19, and 18.

4.7 Geometric mean

The geometric mean is often used to calculate accumulated interest rates on bank accounts. It is the nth root of the product of the n items in a distribution.

> For example,
> The geometric mean of 4 and 9 $= \sqrt{(4 \times 9)} = 6$
>
> The geometric mean of 6, 7 and 8 $= \sqrt[3]{6 \times 7 \times 8} = 6.952$
>
> In general, the geometric mean of $x_1, x_2, \ldots, x_n$ is $\sqrt[n]{x_1 \times x_2 \times \ldots \times x_n}$

Example

One year the interest paid on a bank account is 4%.
The following year the interest paid is 9%.
Use the geometric mean to calculate the equivalent single rate for the two years.

First convert the percentages to decimals:

4% interest means you have 104% of the original balance $= 1.04$

Similarly 9% interest means 109% in total $= 1.09$.

The geometric mean of 1.04 and 1.09 is
$$\sqrt{1.04 \times 1.09}$$
$$= \sqrt{1.1336}$$
$$= 1.0647\ldots$$
So the equivalent single rate is 6.47% (3 s.f.).

Exercise 4G

1. (a) Calculate the geometric mean of
       (i)   3 and 12   (ii) 2, 4 and 8   (iii) 3, 8, 10, 12
   (b) Calculate the arithmetic mean for the above sets and compare
       the two means. What do you notice?

2. The annual rate of interest for two consecutive years was 5% and 7%.
   (a) Calculate the single rate for two years that would pay the same
       amount of interest.
   The third year the interest rate was 4%.
   (b) Calculate the single rate for three years that would pay the
       same amount of interest.

3. The value of a car fell by 30% the first year and by 10% the second
   year. Calculate the single rate of loss for two years which is
   equivalent to the two separate annual losses.

4.
Batsman A	2	18	70
Batsman B	25	26	27

   (a) Calculate the geometric and arithmetic mean for each batsman.
   (b) Which of these two means would batsman A prefer to use in
       comparison with batsman B?
   In their next innings they both score 0.
   (c) Calculate both A and B's new geometric and arithmetic means.

4.8   Range and quartiles

Averages provide a typical value for the data you are looking at. To get the
complete picture you also need to know the spread (or **dispersion**) of data.

Range

Range = largest value − smallest value

The range is a crude measure of spread because it only uses the highest
and lowest values.

Example

The temperature in °C was recorded at 2-hourly intervals at a location in the desert. These are the results:

$$-4, -12, -2, 5, 20, 27, 25, 32, 38, 39, 27$$

(a) Find the median temperature.
(b) Find the range.

(a) Put in order, the numbers are:

$$-12, -4, -2, 5, 20, 25, 27, 27, 32, 38, 39$$

The median is given by the $\frac{1}{2}(11 + 1)$th value, that is the 6th value.

The median temperature is $25\,°C$.

(b) Range = largest value − smallest value

$$= 39 - (-12)$$
$$= 51$$

The range is 51°C.

> The temperature in many deserts can be very hot during the day but very cold at night.
> Therefore the range of temperatures will be a large number.

Quartiles

Quartiles split data up into four equal parts, or quarters. Imagine the data in ascending order of size along a line:

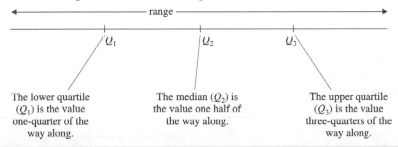

The lower quartile (Q_1) is the value one-quarter of the way along.

The median (Q_2) is the value one half of the way along.

The upper quartile (Q_3) is the value three-quarters of the way along.

> For a list of n data values in ascending order:
>
> ✦ The lower quartile Q_1 is the $\frac{1}{4}(n + 1)$th value.
> ✦ The median Q_2 is the $\frac{1}{2}(n + 1)$th value.
> ✦ The upper quartile Q_3 is the $\frac{3}{4}(n + 1)$th value.

> If $\frac{1}{4}(n + 1)$ is not an integer then round it off to the nearest integer.
> Similarly for $\frac{3}{4}(n + 1)$.

An alternative measure of spread is the **interquartile range**.

> Interquartile range = upper quartile − lower quartile

> The interquartile range gives you the spread of the central 50% of the data.

Example

(a) Find the upper and lower quartiles for the following set of data:

$$7, 2, 9, 11, 4, 5, 12$$

(b) Find the interquartile range for this data.

(a) First order the data   2, 4, 5, 7, 9, 11, 12

$n = 7$

$Q_1 = \frac{1}{4}(7 + 1)$th value, that is the 2nd value.

$Q_1 = 4$

$Q_3 = \frac{3}{4}(7 + 1)$th value, that is the 6th value.

$Q_3 = 11$

(b) Interquartile range $= Q_3 - Q_1$

$= 11 - 4$

$= 7$

Exercise 4H _____

1. Find (i) the range, (ii) the lower quartile, (iii) the upper quartile, and (iv) the interquartile range of the following data sets:

(a) 6, 7, 3, 8, 8, 4, 8, 3, 2, 5, 6, 3, 4, 4, 5, 6, 7, 6, 8, 9, 3, 5, 2
(b) 12, 15, 18, 15, 14, 17, 13, 11, 12, 18, 15, 16, 16, 14, 17, 11, 15, 17, 18
(c) 22, 23, 18, 20, 25, 23, 97, 24, 29, 21, 20, 23, 24, 22, 25, 39, 20

2. Find the median and the upper and lower quartiles from the stem and leaf diagrams you drew in Exercise 3J.

Box and whisker diagrams

Box and whisker diagrams, or box-plots, highlight the quartiles and extreme values of a set of data.

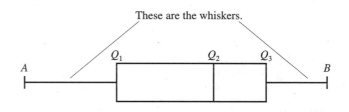
These are the whiskers.

A box-plot looks like this:

A and B are the lowest and highest data values respectively.

> ✦ If Q_2 is closer to Q_1 than to Q_3 then the data is **positively skewed**.
> ✦ If Q_2 is closer to Q_3 than to Q_1 then the data is **negatively skewed**.

A box-plot should have a scale. You can see this in the example:

> Note on **skewness**:
> Skewness describes the **shape** of a distribution. There is more about this on page 76.

● Example

Here are the recorded temperatures (in degrees centigrade) from the first example in this section:

$$-4, -12, -2, 5, 20, 27, 25, 32, 38, 39, 27$$

(a) Find the median and quartiles.
(b) Draw a box-plot to illustrate this data.
(c) Describe the skewness of the distribution.

Put the temperatures in numerical order:

$-12, -4, -2, 5, 20, 25, 27, 32, 38, 39$

The median is $25\,°C$ (see the example on page 105).

There are 11 data values, so $n = 11$.

Lower quartile $Q_1 = \frac{1}{4}(11 + 1)$th value, that is the 3rd value.
$Q_1 = -2\,°C$.

Upper quartile $Q_3 = \frac{3}{4}(11 + 1)$th value, that is the 9th value.
$Q_3 = 32\,°C$.

(b) The box-plot will look like this:

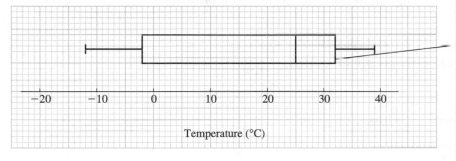

Temperature (°C)

Note: The height of the box is usually 1 or 2 cm.

(c) Q_2 is closer to Q_3 than to Q_1. The distribution is negatively
skewed.

Note: When Q_2 Is closer to Q_1 the distribution is positively skewed.

Exercise 4I

1. For question 1c in the previous exercise, draw a box and whisker diagram.

2. The marks obtained, in a test, by two sets of pupils are given in the following table:

Boys	40	50	46	52	46	51	85
Girls	37	72	39	68	48	74	73

(a) Find the interquartile range of
   (i) the boys' marks,
   (ii) the girls' marks.
(b) Find the range of
   (i) the boys' marks,
   (ii) the girls' marks.
(c) Which set of marks is more dispersed?
(d) Draw box-plots for both the boys' and the girls' marks on the
same diagram. Compare their skewness.

The box-plot in question 2d will look something like this:

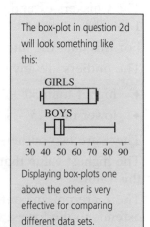

Displaying box-plots one above the other is very effective for comparing different data sets.

Outliers

An outlier may be a value that has been mis-recorded or a value that has been measured and recorded correctly but does not fall in line with the rest of the data.

> Outliers are values that are unusual in comparison with the rest of the data.

To find any outliers you:

✦ find values 1.5 times the interquartile range below the lower quartile and 1.5 times the interquartile range above the upper quartile.

Any values outside this range are outliers and should be marked on your box-plot individually.

Instead of the whiskers showing the full range of values you only draw the whiskers:

✦ up to the highest value that is not an outlier; and
✦ down to the lowest value that is not an outlier.

Example

Show this data using a box and whisker diagram.

20 23 24 24 25 25 26 27 27 28
28 28 29 30 30 30 34 38 40

Find the median and quartiles first:

The quartiles of this data are:

✦ lower quartile 25
✦ median   28
✦ upper quartile 30

These form the values for the box.

The whiskers extend from the quartiles up to a maximum value of 1.5 × the interquartile range.

For this data the interquartile range is 5 ($= 30 - 25$).

The outliers are any values

✦ higher than $30 + 1.5 \times 5 = 37.5$; and
✦ lower than $25 - 1.5 \times 5 = 17.5$

There are two outliers above the upper quartile at 38 and 40.

The highest value that is not an outlier is 34. The whisker extends to this value.

There are not any outliers at the lower quartile and so the whisker extends to the lowest value: 20.

The box and whisker plot looks like this:

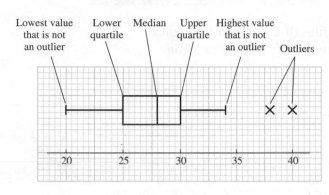

Exercise 4J _____

1. Draw a box-plot for this data:

   59 73 65 86 76
   71 80 72 71 85
   76 72 60 75 70

   Identify any outliers.

2. (a) Draw a box and whisker plot for this data:

       16 23 32 41 44 47
       50 53 54 56 56 58
       58 59 59 60 62 64
       66 67 70 72 73

   (b) Identify any outliers.
   (c) The first listed data item was checked and found to be 61.
       Redraw the box-plot to incorporate this change.

3. Give as many data values as possible for the set of data illustrated
   by the box and whisker plot.
   Describe the skewness of the data.

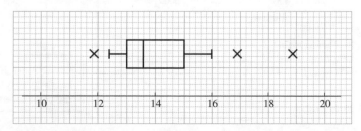

4.9   Measures of spread using a cumulative frequency polygon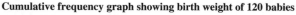

You can find the median and quartiles of a grouped frequency distribution if you first draw a cumulative frequency polygon.

● Example

The table below shows the birth weight (in kilograms) of 120 babies.

Birth weight (kg)	2.0–2.5	2.5–3.0	3.0–3.5	3.5–4.0	4.0–4.5	4.5–5.0
Frequency	12	22	33	27	18	8

For **grouped** data:
Q_1 is the $\frac{1}{4}n$th value
Q_2 is the $\frac{1}{2}n$th value
Q_3 is the $\frac{3}{4}n$th value

(a) Draw a cumulative frequency polygon of this data.
(b) Estimate the median from your graph.
(c) Estimate the interquartile range.
(d) Draw a box and whisker diagram of the birth weights.

(a) The cumulative frequencies are: 12, 34, 67, 94, 112, 120.
So you plot the points (2.5, 12), (3.0, 34), (3.5, 67) and so on and join them with straight lines:

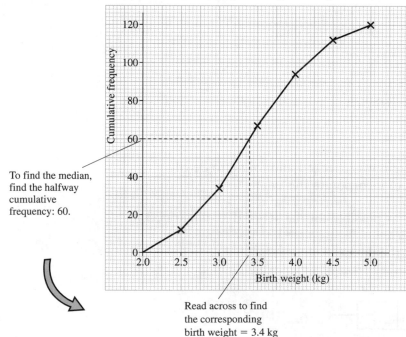

Cumulative frequency graph showing birth weight of 120 babies

To find the median, find the halfway cumulative frequency: 60.

Read across to find the corresponding birth weight = 3.4 kg

You can plot the first point (2.0, 0) as there is no data below 2.0.

The median is a good average to use here as it ignores the outliers.

(b) An estimate of the median is 3.4 kg.

(c) To estimate the interquartile range first find the first and third quartiles on your graph:

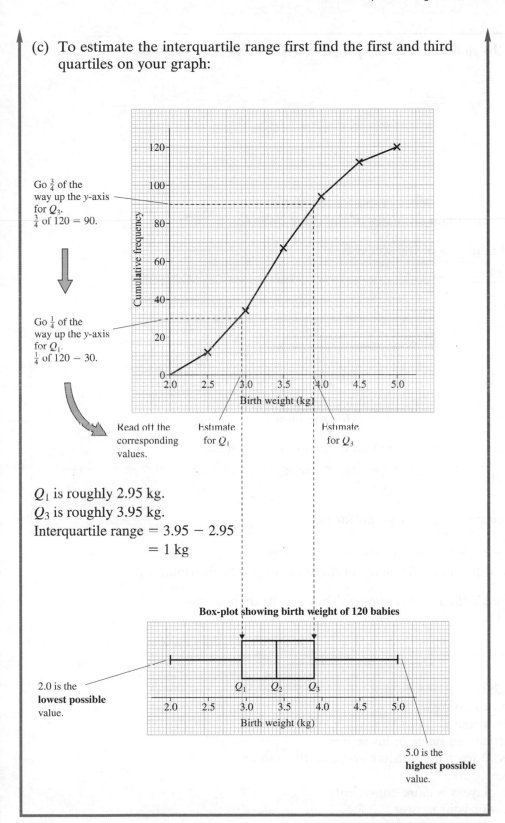

Go $\frac{3}{4}$ of the way up the y-axis for Q_3.
$\frac{3}{4}$ of 120 = 90.

Go $\frac{1}{4}$ of the way up the y-axis for Q_1.
$\frac{1}{4}$ of 120 = 30.

Read off the corresponding values.

Estimate for Q_1

Estimate for Q_3

Q_1 is roughly 2.95 kg.
Q_3 is roughly 3.95 kg.
Interquartile range = 3.95 − 2.95
                    = 1 kg

Box-plot showing birth weight of 120 babies

2.0 is the **lowest possible** value.

5.0 is the **highest possible** value.

Exercise 4K _____

1. The table shows the time taken by 100 people to travel to work.

Time (minutes)	Frequency
10 to 20	9
20 to 25	24
25 to 30	38
30 to 40	15
40 to 60	14

(a) Draw a cumulative frequency graph for this data.
(b) From your graph obtain estimates of the three quartiles.

2. The table gives the heights of 180 students.

Height (nearest cm)	Frequency
140 to 144	36
145 to 149	54
150 to 152	43
153 to 154	22
155 to 164	25

(a) Draw a cumulative frequency graph for this data.
(b) From your graph obtain an estimate for the interquartile range.

3. The age distribution of people living in a certain road is as follows:

Age last birthday	0–9	10–14	15–19	20–29	30–39	40–59	60–79
Frequency	32	25	28	22	35	52	41

> **Hint:**
> The upper class boundary for age is usually a whole number of years.
> Use 10 not 9.5 for the first boundary.

(a) Draw a cumulative frequency graph for this data.
(b) Estimate Q_1, Q_2 and Q_3.
(c) Draw a box-and-whisker plot of the data.
(d) What does your diagram tell you about the skewness of the distribution?

4. Hugo and Boris are brilliant darts players. They record their scores over 60 throws. Here are Hugo's scores.

Score	$20 < x \leqslant 60$	$60 < x \leqslant 90$	$90 < x \leqslant 120$	$120 < x \leqslant 150$	$150 < x \leqslant 180$
Frequency	10	4	13	23	10

(a) Draw a cumulative frequency graph for Hugo's scores.
(b) Use the graph to estimate
  (i) his median score,
  (ii) the interquartile range of his scores.
For his 60 throws, Boris had a median score of 105 and an interquartile range of 20.
(c) Which of the players is more consistent?
   Give a reason for your answer.

Deciles and percentiles

◆ Quartiles divide data into four equal groups.
◆ Percentiles divide data into 100 equal groups.

◆ Deciles divide data into 10 equal groups.

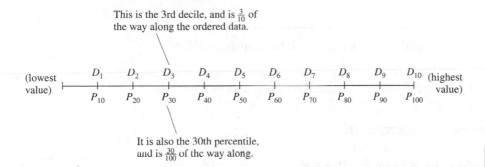

This is the 3rd decile, and is $\frac{3}{10}$ of the way along the ordered data.

It is also the 30th percentile, and is $\frac{30}{100}$ of the way along.

The 10–90 percentile range $= P_{90} - P_{10}$

The 10–90 percentile range differs from the range in that it avoids the extremes, or **outliers** at either end of the data.

Example

Look back at the graph of birth weights.

(a) Estimate the seventh decile.
(b) Estimate the 10–90 percentile range.
   What does this tell you about the central 80% of the data?

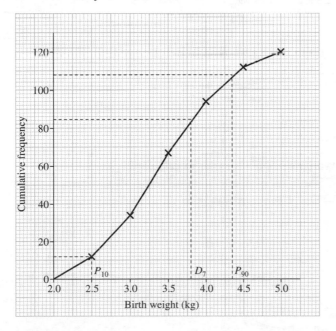

(a) $\frac{7}{10}$ of 120 = 84, so from the graph, $D_7 \simeq 3.8\,\text{kg}$

(b) 10% of 120 = 12 so, $P_{10} \simeq 2.5\,\text{kg}$
$P_{90} \simeq 4.35\,\text{kg}$
10–90 percentile range = 1.85 kg

The middle 80% of birth weights has a range of 1.85 kg.

Exercise 4L ────────────────

1. Estimate the third and seventh deciles for the data in question 1 of Exercise 4K.

2. Estimate the 10–90 percentile range for the data in question 2 of Exercise 4K.

3. Look back at question 3 in Exercise 4K.
    30% of people are above a certain age.
    From your graph, estimate a value for this age.

4.10   Variance and standard deviation

Variance is a measure of spread that uses all the data. The square root of the variance is called the **standard deviation**.
You can use a formula to calculate the variance.

> Note: The interquartile range only uses two values: the upper and lower quartiles. You use it when there are outliers. You use the variance when there are no outliers as it has all the data.

For a set of numbers $x_1, x_2, \ldots, x_n$ with a mean $\bar{x}$:

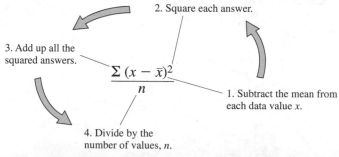

2. Square each answer.

3. Add up all the squared answers.

$$\frac{\Sigma\,(x - \bar{x})^2}{n}$$

1. Subtract the mean from each data value x.

4. Divide by the number of values, n.

There is an alternative formula for variance which is easier to calculate:

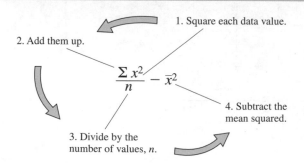

1. Square each data value.

2. Add them up.

$$\frac{\Sigma\,x^2}{n} - \bar{x}^2$$

4. Subtract the mean squared.

3. Divide by the number of values, n.

> This formula is much easier to use when $\bar{x}$ is not a whole number.

The formula for standard deviation is:

$$\sqrt{\frac{\sum (x - \bar{x})^2}{n}} \quad \text{or} \quad \sqrt{\frac{\sum x^2}{n} - \bar{x}^2}$$

Example

For this set of seven numbers: 2, 4, 5, 7, 7, 8 and 9, find the variance and standard deviation

(a) using the formula,

(b) using the alternative formula.

(a) First find the mean:

$$2 + 4 + 5 + 7 + 7 + 8 + 9 = 42$$

$$\bar{x} = 42 \div 7 = 6$$

You can do the calculation for the variance easily if you draw up a table:

x	$x - \bar{x}$	$(x - \bar{x})^2$
2	4	16
4	−2	4
5	−1	1
7	1	1
7	1	1
8	2	4
9	3	9
Total		36

This is $\sum (x - \bar{x})^2$

Variance $= 36 \div 7 = 5.14\ldots$

Standard deviation $= \sqrt{5.14\ldots}$

$$= 2.27 \text{ (to 2 d.p.)}$$

(b) Square the data and add them up: $\sum x^2 = 4 + 16 + 25 + 49 + 49 + 64 + 81$

$$= 288$$

Divide by the number of values: $288 \div 7 = 41.14\ldots$

Subtract the squared mean: $41.14\ldots - 6^2 = 5.14\ldots$

Standard deviation $= \sqrt{5.14\ldots}$

$$= 2.27 \text{ as before}$$

Exercise 4M

1. Calculate the mean and standard deviation of the following data sets. Use the formula $\sqrt{\dfrac{\sum(x-\bar{x})^2}{n}}$ for the standard deviation.

   (a) 4, 6, 7, 7, 8, 10, 10, 7, 5, 4, 4, 2, 3, 6, 7
   (b) 3, 6, 9, 12, 13, 15, 17, 19, 23, 28, 31
   (c) 8, 12, 5, 8, 13, 5, 2, 16, 4, 15, 6, 6, 17

2. Calculate the standard deviation for the data in question 1 using the alternative formula.

   Hint:
   Check that your answer is the same as for question 1.

3. (a) Calculate the mean and standard deviation of the following data sets using whichever formula you prefer.
   (i)   6, 7, 3, 2, 7, 5, 11, 4, 7, 6, 1, 3, 8, 3, 2
   (ii) 60, 70, 30, 20, 70, 50, 110, 40, 70, 60, 10, 30, 80, 30, 20
   (iii) 4, 5, 1, 0, 5, 3, 9, 2, 5, 4, −1, 1, 6, 1, 0

   (b) What do you notice about the mean and standard deviation of the data sets in part (a)?

Standard deviation for a frequency distribution

p259

p260

For a discrete frequency distribution, the standard deviation is

$$\sqrt{\dfrac{\sum f(x-\bar{x})^2}{\sum f}} \quad \text{or} \quad \sqrt{\dfrac{\sum fx^2}{\sum f} - \bar{x}^2}$$

Note:
The formula is very similar to the formula for a discrete set of numbers but $\sum f$ is used instead of n.

Example

The data below gives the number of books read in the last month by a class of 20 students.

Number of books, x	0	1	2	3	4
Number of students, f	2	5	6	5	2

(a) Find the mean and standard deviation of the number of books. Use the formula $\sqrt{\dfrac{\sum f(x-\bar{x})^2}{\sum f}}$ for the standard deviation.

(b) Use the alternative formula for the standard deviation to check your answer.

(a) First set out a table like this:

x	f	fx	$x - \bar{x}$	$(x - \bar{x})^2$	$f(x - \bar{x})^2$
0	2	0	-2	4	8
1	5	5	-1	1	5
2	6	12	0	0	0
3	5	15	1	1	5
4	2	8	2	4	8
Totals	20	40			26

$\bar{x} = 40 \div 20 = 2$

Std. dev. $= \sqrt{\frac{26}{20}} = 1.14$ books

> Standard deviation is in the same units as the data

(b) $\sum fx^2 = 2 \times 0^2 + 5 \times 1^2 + 6 \times 2^2 + 5 \times 3^2 + 2 \times 4^2$

$\qquad = 106$

$\dfrac{\sum fx^2}{\sum f} = 5.3$

$\dfrac{\sum fx^2}{\sum f} - \bar{x}^2 = 5.3 - 2^2$

$\qquad\qquad = 1.3$

So the variance is 1.3.

Std. dev. $= \sqrt{1.3} = 1.14$ books.

> For a grouped frequency distribution the formula is exactly the same as for a discrete distribution. You use the **mid-interval value** for x.

Example

The table shows the time taken by a group of 25 students to solve a problem.

Find the mean and standard deviation of this data.

Use the formula $\sqrt{\dfrac{\sum fx^2}{\sum f} - \bar{x}^2}$ for the standard deviation.

Time (to the nearest second)	10–14	15–19	20–24	25–29
Frequency (f)	3	5	10	7

Redraw the table:

Time (s)	Frequency, f	Mid-value x	fx	fx^2
10–14	3	12	36	432
15–19	5	17	85	1445
20–24	10	22	220	4840
25–29	7	27	189	5103
Totals	25		530	11 820

> This is the middle value in each interval.

Mean $\bar{x} = \dfrac{\sum fx}{\sum f} = \dfrac{530}{25} = 21.2$ seconds

Standard deviation $= \sqrt{\left(\frac{11\,820}{25} - 21.2^2\right)} = 4.83$ seconds (to 3 s.f.)

Exercise 4N

In questions 1 to 4, calculate the mean and standard deviation using the values given:

1. $\sum (x - \bar{x})^2 = 156$ $\sum x = 53$ $n = 10$

2. $\sum x^2 = 247$ $\sum x = 40$ $n = 8$

3. $\sum (x - \bar{x})^2 = 293$ $\sum x = 98$ $n = 12$

4. $\sum x^2 = 941$ $\sum x = 84$ $n = 8$

> **Remember** for questions **1** to **4**:
>
> mean $\dfrac{\sum x}{n}$
>
> standard deviation =
>
> $\sqrt{\dfrac{\sum (x - \bar{x})^2}{n}}$
>
> or $\sqrt{\dfrac{\sum x^2}{n} - \bar{x}^2}$

5. Calculate an estimate of the mean and standard deviation for this data showing the masses of various objects.

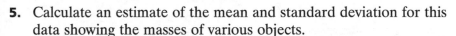

Mass (kg)	$10 < m \leqslant 20$	$20 < m \leqslant 30$	$30 < m \leqslant 40$	$40 < m \leqslant 50$
Frequency	12	23	43	9

6. The table gives the lengths of words in a crossword puzzle. Calculate the mean and standard deviation.

Length of word	4	5	6	7	8	9	10	11
Frequency	5	8	9	4	5	8	4	2

7. The number of matches in a box was counted for a sample of 40 boxes. The results are summarized in the table.

Number of matches	31	32	33	34	35	36	37
Frequency	1	5	6	7	8	7	6

Calculate the mean and the standard deviation of the number of matches in a box.

8. Calculate an estimate of the mean and standard deviation of the times taken by a number of people to solve a problem.

Time (nearest second)	10 to 14	15 to 19	20 to 24	25 to 29	30 to 34
Frequency	9	15	16	12	8

4.11 Standardized scores

Standardized scores allow you to compare values from different data sets. You need to know the mean and standard deviation.

Standardized score $z = \dfrac{\text{score} - \text{mean}}{\text{standard deviation}}$

Example

Louise and David's Maths and English marks are shown in the table below along with the mean and standard deviation for the whole class.

	Louise's mark	David's mark	Mean mark	Standard deviation
Maths	75	50	60	10
English	65	40	50	5

Work out Louise and David's standardized scores and comment on their performance.

Louise's standardized English score can be worked out like this:

$$z = \frac{\text{score} - \text{mean}}{\text{standard deviation}}$$

$$z = \frac{65 - 50}{5}$$

$$= 3$$

The standardized scores can be shown in a table:

Standardized score

	Louise	David
Maths	1.5	−1
English	3	−2

Louise did better than David overall.

The 'best' mark is Louise's English mark of 65 because it has the highest standardized score, while the 'worst' mark is David's English mark of 40.

> **Note:**
> Standardized scores greater than zero are above average for the data set. Standardized scores less than zero are below average.
> A standardized score of zero is average.

Exercise 40

1. The table shows Carla's height and weight, alongside the mean and standard deviation for her class.

	Carla's values	Class mean	Class standard deviation
Weight (kg)	48	44	8
Height (cm)	160	175	10

Calculate the standardized scores for Carla's height and weight.

2. The table gives exam results for three students, with the mean and standard deviations for their whole class.

	Lisa	Mark	Pierre	Class mean	Class standard deviation
French	70	55	67	60	5
German	69	47	66	55	8

(a) Calculate the six standardized scores.
(b) Who did equally well in the two exams?

4.12 The normal distribution

The Normal distribution is a perfectly symmetrical bell-shaped curve:

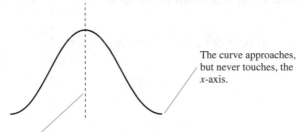

The curve approaches, but never touches, the *x*-axis.

Real life examples include:
Heights of a random sample of people.
The precise amount of cola in a 330 ml can.

The line of symmetry shows the average of the distribution. It is where the mean, median and mode coincide.

It is called a 'Normal' distribution because it has few outlying values.

> Approximately 95% of the distribution is within 2 standard deviations of the mean and over 99% is within 3 standard deviations.

The diagrams on the next page show the position and spread for three normal distributions. For each the total relative frequency is the same. The three are

mean 0 and standard deviation 1   ———
mean 2 and standard deviation 1   ···············
mean 0 and standard deviation 2   – – – –

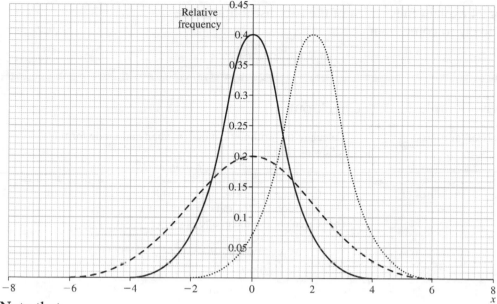

Normal curves

Note that
+ the curve is symmetrical about the mean.
+ if the standard deviation is increased then the curve becomes less peaked and more spread out.

Exercise 4P

1. Sketch on the same axes normal distributions with

	A	B	C
Mean	20	25	16
Standard deviation	4	6	8

2. Sketch on the same axes normal distribution with

	P	Q
Mean	100	90
Standard deviation	10	20

It is not known if a particular value comes from distribution P or distribution Q.
Use your sketches to explain why a value of 130 is more likely to come from Q even though it is nearer to the mean of P.

3. Sketch on the same axes normal distributions with

	R	S
Mean	28	26
Standard deviation	3	2

It is not known if a particular value comes from distribution R or distribution S.
For a value of 24 explain fully which of the two distributions the value is most likely to come from.

Summary

You should now be able to	Check out 4
1 Calculate a variety of 'averages'.	**1** Calculate the arithmetic mean, median and mode for the following data 5, 2, 7, 6, 5, 7, 7
2 Calculate a variety of measures of spread.	**2** (a) Calculate the range, interquartile range and standard deviation for the data in question 1 (b) Calculate the mean and standard deviation for this data.

Length (cm)	Frequency
$20 \leqslant L < 40$	24
$40 \leqslant L < 50$	28
$50 \leqslant L < 60$	30
$60 \leqslant L < 80$	36
$80 \leqslant L < 110$	24

(c) Draw a cumulative frequency graph to obtain estimates of the median and interquartile range of the data in (b).

| **3** Use standardized scores. | **3** The table shows some of the exams taken by Amit with the mean and standard deviation of his class. |

	Mean	Standard deviation	Amit's result
History	55	5	65
Geography	60	10	70

Amit's history teacher claims that he has done better in History than in Geography. Justify his teacher's claim.

| **4** State the advantages and disadvantages of each measure. | **4** Which measure of average would you calculate for this set of data?
4, 4, 25, 2, 4, 1, 3, 2, 2
Give a reason for your choice. |
| **5** Use the normal distribution. | **5** X is distributed normally with mean 50 and standard deviation 10. Y is distributed normally with mean 60 and standard deviation 20.
Sketch the two distributions. |

Revision Exercise 4

1. The weekly wages (in pounds) of the twenty workers in a factory
are shown below.

85   90   90   90   85   85   120   85   85   85
90   120   160   220   85   90   120   335   160   120

(a) Copy and complete the frequency
distribution.
(b) The shop steward says that the
average wage of the workers is £85.
Which average is he referring to,
the mean, the median or the mode?

Wage (in £)	Tally	Frequency
85		
90		
120		
160		
220		
335		

[NEAB]

2. The number of goals scored by the 11 members of a hockey team
in 1993 were as follows:

6   0   8   12   2   1   2   9   1   0   11

(a) Find the median.
(b) Find the upper and lower quartiles.
(c) Find the interquartile range.
(d) Explain why, for this data, the interquartile range is a more
appropriate measure for spread than the range.
(e) The goals scored by the 11 members of the hockey team in
1994 are summarized in the box-plot below.

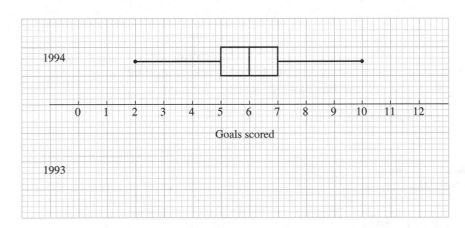

(i) Copy the diagram above and summarize the results for
1993 in the same way.
(ii) Do you think the team scored more goals in 1994?
Explain your reasoning.   [NEAB]

3. A sample of 27 commuters were asked to record the number of times they travelled to work by rail over a six-month period.

The numbers recorded were as follows.

48	40	59	54	71	60	42
35	42	63	61	54	69	81
36	47	50	46	73	71	64
28	35	45	63	75	56	

(a) Construct a stem and leaf diagram of this information, ensuring that the figures on the leaves are in order of size.

(b) Use your diagram to find
  (i)   the median,
  (ii) the interquartile range.   [NEAB]

4. A lake contains an unknown number of fish. A random sample of 50 fish is taken from the lake and marked with a red dye before being returned to the water. Twenty samples of size 25 are then taken **with replacement** from the lake and the number of fish per sample having a red mark is noted as follows.

0	2	0	5	2	3	1	4	0	2
1	3	2	1	1	2	3	2	2	1

(a) Use these results to estimate the total number of fish in the lake.

(b) State **two** conditions which must be met to ensure the validity of the estimation method used in part (a).   [NEAB]

5. The histogram opposite shows the alcohol consumption for a random sample of 100 adult males in the UK. The results show consumption in units of alcohol and relate to the week preceding the interview which formed the basis of the investigation.

(a) Copy and complete the following frequency table for these data.

Consumption units of alcohol (x)	Frequency
$0 < x \leqslant 5$	
$5 < x \leqslant 10$	
$10 < x \leqslant 20$	
$20 < x \leqslant 35$	
$35 < x \leqslant 50$	

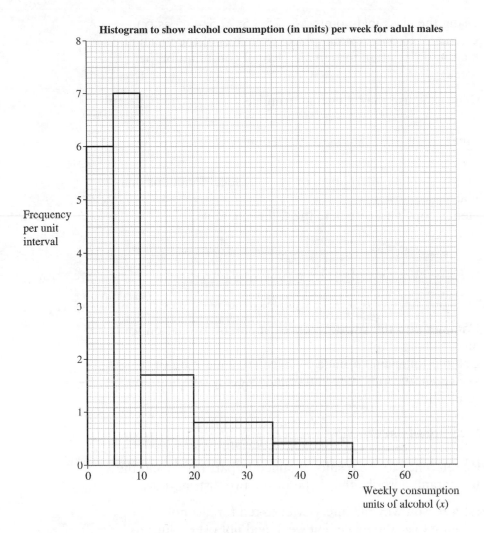

Histogram to show alcohol comsumption (in units) per week for adult males

Frequency per unit interval

Weekly consumption units of alcohol (x)

(b) Equivalent data for a sample of 100 females was:

Consumption units of alcohol (x)	Frequency
$0 < x \leqslant 5$	77
$5 < x \leqslant 10$	10
$10 < x \leqslant 20$	9
$20 < x \leqslant 35$	3
$35 < x \leqslant 50$	1

(i)   Using the class mid-points 2.5, 7.5, 15, 27.5, 42.5 as (x), calculate, to one decimal place, estimates of the mean and standard deviation for female alcohol consumption per week.

(You may use the following formulae

$$\text{Mean} = \frac{\sum fx}{\sum f} \qquad \text{Standard deviation} = \sqrt{\frac{\sum f(x - \bar{x})^2}{\sum f}}$$

or any suitable alternative, or the statistical functions on your calculator.)

(ii) Hence copy and complete the following table:

	Median	Mean	Standard deviation
Male	7.9	11.8	11.0
Female	3.3		

Comment on the key differences apparent in relation to alcohol consumption rates.

(c) (i)   Use the following formula to calculate a measure of skewness for both male and female weekly consumption of alcohol.

$$\text{Measure of skewness} = \frac{3\,(\text{mean} - \text{median})}{\text{standard deviation}}$$

(ii) By reference to the results obtained in part (c) (i) and the histogram, describe the shape of the two distributions.

(d) Suggest a reason why the interviewer asked for alcohol consumption over the preceding week and not over a longer period.

(e) Indicate how the survey should be continued if the results are to be representative of changes in alcohol consumption throughout the year.   [NEAB]

6. A hospital recorded the birth weights, in kilograms, of 100 girls and 100 boys. The weights are summarized in the table below.

	Median	Lower quartile	Upper quartile	Minimum	Maximum
Girls	3.1	2.5	3.7	1.3	4.5
Boys	3.3	2.4	4.0	1.0	4.8

(a) Draw **two** box-plots for the birth weights of girls and boys.
(b) Use the information to write a brief comparison of the birth weights of the girls and boys.   [NEAB]

7. The diagram below shows the cumulative distribution of times, in minutes, of the lengths of a sample of 100 telephone calls made from a City office.

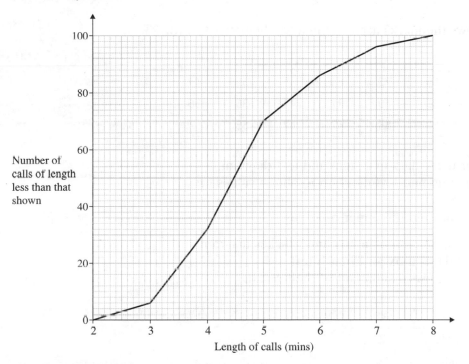

Number of calls of length less than that shown

Length of calls (mins)

Use the diagram to:

(a) Find the number of calls of less than 6 minutes length.
(b) Estimate the median length of calls made.
(c) Estimate the upper and lower quartiles of the data.
(d) Estimate the time exceeded by 60% of the calls.
(c) Find the probability that any two of these one hundred calls chosen at random each exceeded 5 minutes in length.   [NEAB]

8. The length of 300 telephone calls from ordinary telephones is given in the table.

Time T (seconds)	$0 \leqslant T < 40$	$40 \leqslant T < 60$	$60 \leqslant T < 80$	$80 \leqslant T < 100$	$100 \leqslant T < 120$	$120 \leqslant T < 160$	$160 \leqslant T < 200$
Frequency	38	36	41	58	49	48	30

(a) Draw a histogram of these data.

(b) Calculate an estimate of the mean and standard deviation of the length of calls.

The mean and standard deviation of the length of calls from mobile telephones are 64 seconds and 35 seconds respectively.

(c) Comment on the differences in the length of calls from mobile and ordinary telephones.

9. The table shows the marks achieved in a Mathematics test for 23 candidates.

43	51	65	73	84	55	39	44
23	78	43	49	27	44	36	56
54	48	56	55	67	67	55	

(a) Represent these data in a stem and leaf diagram.
(b) Use your stem and leaf diagram to obtain values for
   (i) the median,
   (ii) the lower and the upper quartiles.

(c) Represent these data on a box and whisker diagram.
The same candidates took an English test.
The table below shows a summary of their results.

Lowest mark	32
Lower quartile	49
Median mark	56
Upper quartile	63
Highest mark	71

(d) Make **two** comments comparing the Mathematics and English test results.   [SEG]

10. The bills for a household in January 1999 are shown below.

Gas	Electricity	Council tax	Telephone
£101	£72	£95	£60

The total of the four bills was £328.

(a) Draw a pie chart to illustrate these data.

The bills for January 2000 are shown below.

Gas	Electricity	Council tax	Telephone
£108	£75	£102	£72

The diameter of the pie chart for the year 1999 is 9.6 cm.

(b) Calculate the diameter of the comparative pie chart for the year 2000.

11. The cumulative frequency curve represents the times taken to run 1500 metres by each of the 240 members of the athletics club, Weston Harriers.

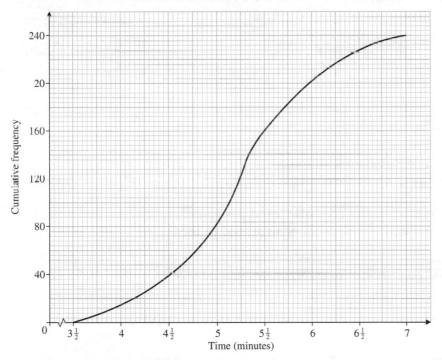

(a) From the graph, find
   (i)   the median time,
   (ii) the upper quartile and the lower quartile.

(b) Draw a box and whisker diagram to illustrate the data.
(c) Use your box and whisker diagram to make **one** comment about the shape of a histogram for these data.

A rival athletics club, Eastham Runners, also has 240 members. The time taken by each member to run 1500 metres is recorded and these data are shown in the following box and whisker diagram.

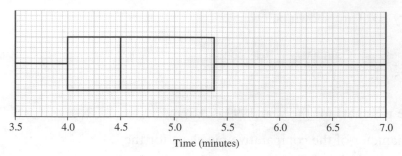

Time (minutes)

(a) Use this diagram to make **one** comment about the data for Eastham Runners as compared with that for Weston Harriers.

[SEG]

12. The table shows the number of goals scored by Debden Dockers in 50 hockey games.

Goals	Frequency
0	4
1	9
2	20
3	11
4	6
Total	50

(a) Calculate the mean number of goals scored per game.
(b) Calculate the standard deviation of goals scored in these games.

Loughton Farmers' goal record for its 50 games had a bigger standard deviation, but a smaller range.

(c) Explain how this can occur. [SEG]

13. The table shows the distribution of the monthly income of 120 factory workers.

Income (£)	Frequency
400 and under 420	12
420 and under 440	27
440 and under 460	34
460 and under 480	24
480 and under 500	15
500 and under 520	8

(a) Copy and complete the cumulative frequency distribution table.

Income (£)	Cumulative frequency
under 400	0
under 420	

The diagram shows the cumulative frequency curve for a group of 120 farmers.

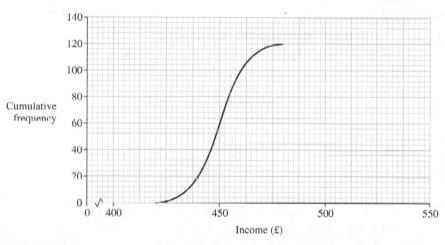

Income (£)

(b) Copy the cumulative frequency curve for farmers and draw the cumulative frequency curve for the 120 factory workers on the same axes.

(c) Use your graph to estimate the interquartile range for the factory workers.

(d) From your graph,
   (i) name the statistical property which is approximately the same for both distributions.
   (ii) name one statistical difference between the distributions.

[SEG]

14. The number of flower buds on a sample of 10 azalea plants were recorded.

   25   37   28   16   37   50   48   45   42   49

(a) Calculate the range of these data.

(b) Calculate the standard deviation of these data.

You may use the statistical functions on your calculator or the following formula.

$$\text{Standard deviation} = \sqrt{\frac{\sum x^2}{n} - \left\{\frac{\sum x}{n}\right\}^2}$$

The number of flower buds on the same 10 plants were recorded two years later. The second set of data had an increased range of 35 but a reduced standard deviation of 7.9.

(c) Write down two differences which you would expect to see in the values of the second set of data as compared with those of the first set of data.

Another type of plant had a positively skewed distribution with a mean of 50 buds per plant.

(d) Sketch this skewed distribution, clearly indicating possible positions of the mean, mode and median.   [SEG]

15. The diagram represents the response times to the ringing of the telephone in an office.

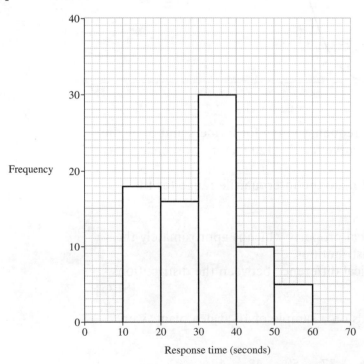

(a) Use the diagram to construct the cumulative frequency table for these response times.

(b) Draw a cumulative frequency polygon.

(c) Obtain an estimate for the interquartile range of these response times.

Six months later a similar survey of response times gave an interquartile range of 10 seconds.

(d) What does this new value suggest about the change in the distribution of response times at this office?   [SEG]

16.

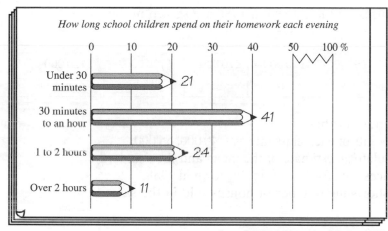

How long school children spend on their homework each evening

Source: Which Magazine October 1993

The above information was part of a survey, by questionnaire, given to 400 pupils.

(a) Calculate the number of pupils who said that they worked more than two hours every evening.

(b) What percentage of the pupils did not answer this part of the questionnaire?

(c) Give a suitable reason why these children may not have recorded an answer to this part of the questionnaire.   [SEG]

17. (a) A machine is used to pack paper clips. In order to test the efficiency of the machine, twenty packets were taken at random and the contents counted.

Number of paper clips in a packet	Number of packets		
48	2		
49	2		
50	6		
51	5		
52	0		
53	1		
54	2		
55	2		

Calculate

(i)   the mean number of paper clips in each packet,
(ii) the standard deviation.

(b) Another machine packs paper clips with a mean 51 and
standard deviation 4.2. By comparing the machines give
reasons why you would prefer one to the other for efficiency.

[SEG]

18. The prices paid for houses sold in a town in 1988 are shown in the
table below.

House price	£30 000–	£50 000–	£60 000–	£70 000–	£100 000–	£120 000–£150 000
Number sold	2	9	14	51	8	6

(a) Construct a histogram for these data.
(b) By taking the midpoints of each class interval: 40 000, 55 000,
65 000, ... etc, calculate an estimate of the mean and the standard
deviation of the prices paid for houses in this town in 1988.

The histogram below shows the number of houses sold in the same
town in 1991.

(c) How many houses were sold in 1991?
(d) The mean and standard deviation of the prices paid for houses
in 1991 were £70 000 and £21 200 respectively.
(i)   In which year, 1988 or 1991, did the greater variation in
sale prices occur?
(ii) Calculate the difference in the mean sale price of houses in
1988 and in 1991 as a percentage of the 1988 mean sale price.
(e) Give one reason why houses may not be cheaper in 1991 than
in 1988. [SEG]

19. The histogram represents the fish caught by an angler during 1986.

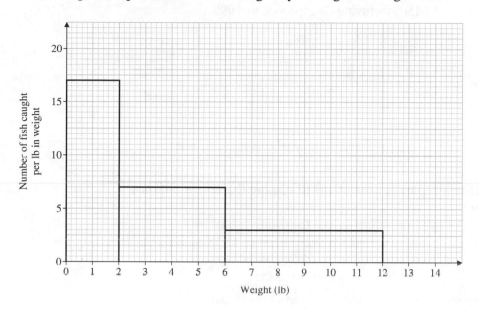

(a) (i) How many fish less than 2 lb in weight did he catch?
   (ii) How many fish did he catch altogether?
        Show clearly how you obtain your answer.

(b) Estimate the total weight of his catch for 1986.   [SEG]

20. A sample of 27 commuters were asked to record the number of times they travelled to work by rail over a six-month period.

The numbers recorded were as follows.

48	40	59	54	71	60	42
35	42	63	61	54	69	81
36	47	50	46	73	71	64
28	35	45	63	75	56	

(a) Construct a stem and leaf diagram of this information, ensuring that the figures on the leaves are in order of size.
(b) Use your diagram to find
   (i) the median,
   (ii) the interquartile range.
(c) Construct a box-plot of this data.
(d) One additional commuter recorded the number of times she travelled by rail over the six-month period as 7.
   (i) By reference to the box-plot, how would you classify this figure?
   (ii) Explain your answer.   [NEAB]

21. The following table shows the profits (or losses) of 100 Retailing Companies and 100 Manufacturing Companies in 1999.

Profits (£000's)	Retailing	Manufacturing
$-10 \leqslant x < 0$	2	12
$0 \leqslant x < 10$	5	25
$10 \leqslant x < 20$	12	35
$20 \leqslant x < 30$	20	18
$30 \leqslant x < 40$	61	10

(a) Construct **two** separate histograms to display the above data for each sector.

(b) Calculate the missing values in the following table.

	Retailing	Manufacturing
Mean profit		£14 400
Median profit	£33 606	£13 714
Standard deviation		£12 375

(c) A measure of skewness is given by the expression

$$3 \times \frac{\text{(mean} - \text{median)}}{\text{standard deviation}}.$$

(i) Use the results from your completed table in part (b) to calculate to 2 decimal places a measure of skewness for each sector.

(ii) What do the measures of skewness and the histograms drawn in part (a) indicate about the shape of the two distributions   [NEAB]

22. In an attempt to devise an aptitude test for applicants seeking work on a factory's assembly line, it was proposed to use a simple construction puzzle. The times taken to complete the task by a random sample of 90 employees were observed with the following results.

Times to complete the puzzle (seconds) x	Number of employees	Cumulative frequency
$10 \leqslant x < 20$	5	5
$20 \leqslant x < 30$	11	
$30 \leqslant x < 40$	16	
$40 \leqslant x < 45$	19	
$45 \leqslant x < 50$	14	
$50 \leqslant x < 60$	12	
$60 \leqslant x < 70$	9	
$70 \leqslant x < 80$	4	

(a) Copy and complete the table.
(b) Construct a cumulative frequency polygon for this data.
(c) Identify from this polygon:
   (i)   the median (Q_2),
   (ii) the upper quartile (Q_3),
   (iii) the lower quartile (Q_1).
(d) (i)   Calculate a value for ($Q_3 - Q_2$) and for ($Q_2 - Q_1$).
   (ii) What do these show about the shape of the distribution?
(e) It is decided to grade the applicants on the basis of their times taken, as good, average or poor.
   The percentages of applicants in these grades are to be approximately 15%, 70% and 15%, respectively.

   Estimate, from your cumulative frequency polygon, the grade limits.   [NEAB]

23. The number of traffic accidents recorded per day over a 200-day period at a busy road junction in the town of Fernlea were as follows.

Number of accidents per day x	Number of days
0	24
1	88
2	61
3	20
4	5
5	2
	200

(a) Calculate for this data
   (i)   the mean ($\bar{x}$),
   (ii) the standard deviation (s), to 2 decimal places.
      (You may use the following formulae

$$\text{Mean} = \frac{\sum fx}{\sum f}$$

$$\text{Standard deviation} = \sqrt{\frac{\sum fx^2}{\sum f} - \bar{x}^2} \text{ or } \sqrt{\frac{\sum f(x - \bar{x})^2}{\sum f}}$$

   or any suitable alternative, or the statistical functions on your calculator.)

An earlier study conducted over a similar time period at the same road junction produced the following summarized results.

Daily road accidents	
Mean ($\bar{x}$)	Standard deviation (s)
3.2	1.15

A comparison of the relative variation present in each of the two sets of data is to be undertaken using the following measure:

$$\frac{s}{\bar{x}}$$

(b) Calculate, to 2 decimal places, this measure for **each** data set.

(c) Comment on the results which you obtained in parts (a) and (b).   [NEAB]

24. The table below shows the distribution of wages of the 11 workers in a small business.

Director	£54 000
Manager	£40 000
Salesman	£22 000
Foreman	£18 000
Welder	£17 000
Patternmaker	£15 000
1st Metalworker	£12 000
2nd Metalworker	£12 000
3rd Metalworker	£12 000
4th Metalworker	£12 000
Apprentice	£6 000

(a) For distribution of wages find the
    (i)   median,
    (ii) lower quartile,
    (iii) upper quartile,
    (iv) interquartile range,
    (v)   mean.

(b) The director was given a wage rise of £22 000.
    Find the **new** values for the
    (i)   median,
    (ii) lower quartile,
    (iii) upper quartile,
    (iv) interquartile range,
    (v)   mean.   [NEAB]

25. The table shows the total
number of goals scored
in 165 hockey matches.
(a) Draw a cumulative **step**
polygon for these data.
(b) Use your graph to find:
    (i) the second decile,
    (ii) the eighth decile.
(c) How many games
resulted in a total score
completely within the
middle 60% of the
total scores?
              [SEG]

Number of goals	Number of matches
0	10
1	15
2	21
3	35
4	25
5	20
6	16
7	11
8	12

26. The table shows the mean and the standard deviation of the heights
of a sample of adult males and a sample of boys aged nine.
The heights of both the adult males and the boys are normally
distributed.

	Adult males	Boys aged nine
Mean	180 cm	135 cm
Standard deviation	18 cm	10 cm

David is a boy aged nine whose height is 120 cm.

(a) How many standard deviations below the mean is David's
height?

It is believed that the height of a boy aged nine is a good indicator
of his adult height.

(b) Estimate the height that David will be when he is an adult.

The diagram shows the distribution of the heights of the boys aged
nine.

(c) Copy the diagram and sketch the distribution of the heights of
the adult males.

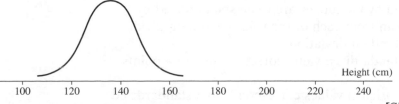

Height (cm)

100   120   140   160   180   200   220   240

[SEG]

27. A class of students is given a History test and Physics test.

Both the History and Physics marks are approximately normally distributed.

The mean and the standard deviation of each distribution are shown in the table.

	Mean	Standard deviation
History	52	6
Physics	60	8

The graph shows a sketch of the distribution for the History marks.

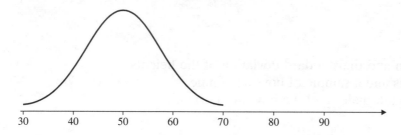

(a) Show, on a copy of the same graph, a sketch of the distribution of the Physics marks.

Kelly scores 64 in the History test and 72 in the Physics test, but she claims that she is better at History than at Physics.

(b) By standardizing her marks, find out whether her test results support her claim. [SEG]

28. The points awarded (out of a maximum of 6.0) by the 12 judges at a recent ice skating tournament were:

5.7, 5.9, 5.9, 5.7, 3.0, 5.8, 5.7, 5.7, 5.5, 5.8, 5.8, 5.5

(a) Calculate the mean points scored.

(b) Using the formula $\sqrt{\dfrac{\sum (x - \bar{x})^2}{n}}$ or the statistical functions on your calculator, or otherwise, calculate to two decimal places the standard deviation of the points scored.

(c) The points awarded by the judges are to be standardized by subtracting the mean from each of the values and then dividing the result by the standard deviation.
   (i) What is the standardized value corresponding to a points score of 5.8?
   (ii) Which of the original values corresponds to a standardized score of +0.263? [NEAB]

29. The time taken, in seconds, by 120 students to complete a puzzle is given in the following table.

Time (s)	Frequency
$20 \leqslant t < 30$	38
$30 \leqslant t < 40$	44
$40 \leqslant t < 60$	20
$60 \leqslant t < 80$	14
$80 \leqslant t < 100$	4

(a) Draw a cumulative frequency graph of the data.
(b) Use the cumulative frequency graph to obtain estimates of:
 (i) the median,
 (ii) the interquartile range.
(c) Calculate an estimate of the mean and standard deviation of the times.
(d) Why, in this case, are the median and interquartile range better measures than the mean and standard deviation?   [SEG]

30. The mean and standard deviation of two physics papers are shown. The marks for both papers are normally distributed.

	Mean	Standard deviation
Paper 1	45	8
Paper 2	56	12

Sarah was present for paper 1 but absent for paper 2.
In paper 1 she scored 59 marks.

(a) What would be the best estimate for her mark in paper 2?
(b) On a copy of the diagram sketch the distributions for both paper 1 and paper 2.

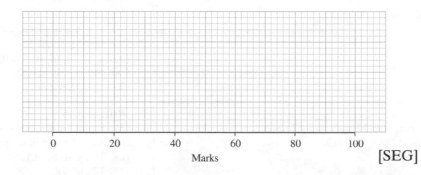

Marks   [SEG]

5 Other summary statistics

> *He uses statistics as a drunken man uses lamp-posts*
> *– for support rather than illumination*
> Andrew Lang

Charts and graphs can be used to show changes over time:

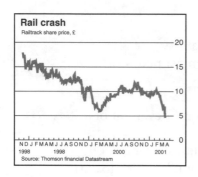

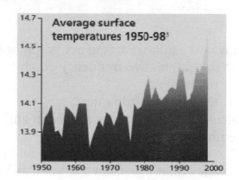

This unit will show you how to

+ Calculate weighted averages
+ Find index numbers
+ Calculate population averages
+ Draw time series graphs
+ Consider quality assurance

Before you start

You need to know how to	Check in 5
1 Find the mean Remember: mean $= \dfrac{\Sigma \text{ values}}{\text{no. of values}}$ Look back at page 94 for more information.	**1** Find the mean of (a) 6, 18, 37, 14 and 22 (b) 29, 36, 18, 22, 13 and 14
2 Plot points on a graph	**2** Plot these points: (5, 3), (2, 3), (2, 6), (5, 6). Join the points in order. What shape do they make?
3 Use ratio In a box of chocolates there are 12 milk and 17 plain. The ratio milk : plain chocolates = 12 : 17.	**3** Write as ratios: (a) 8 girls to 15 boys (b) 30 red to 13 black pens (c) 19 ice-lollies to 12 choc-ices

5.1   Weighted averages

In your GCSE Statistics examination, coursework accounts for a quarter (0.25) of the overall mark and the written paper for three-quarters (0.75) of the overall mark.

The two components contribute differently to the overall mark. To find an overall mark you need to weight the components of the average according to the contribution they make. This is a weighted average.

$$\text{Weighted average} = \frac{\Sigma\,(\text{value} \times \text{weight})}{\Sigma\,(\text{weights})}$$

Remember:
Σ means 'the sum of'.

Example

Lucy achieved 45% for her coursework and 62% in the written exam paper. Using the weights 0.25 for coursework and 0.75 for the written paper, find her overall mark.

First put her results in a table so you can see the information more clearly:

	Coursework	Written paper
% Mark (value)	45	62
Weight	0.25	0.75
Value × weight	45 × 0.25 = 11.25	62 × 0.75 = 46.5

$\Sigma\,(\text{Value} \times \text{weight}) = 11.25 + 46.5 = 57.75$

$\Sigma\,\text{Weights} = 0.25 + 0.75 = 1$

Lucy's overall % mark $= \dfrac{57.75}{1}$

$= 57.75\%$

Note that the weighted average is a percentage because the marks were given as percentages.

Exercise 5A

1. In an examination, Paper 1 is worth 0.2 of the total marks, Paper 2 is worth 0.55 of the marks and coursework is worth 0.25.
Andrew scored 80% on Paper 1, 62% on Paper 2 and 55% for his coursework.
Bethany scored 64% in Paper 1, 68% on Paper 2 and 65% for her coursework.
Who got the highest score overall?

2. The following table gives Richard's % marks and their respective weights in his music exam.

	Performance	Theory	Composition
% Mark	68	72	88
Weight	50	30	20

    An overall mark of 75% would give Richard a distinction for the exam. What percentage did Richard get? Did he gain a distinction?

3. There are four parts in an English test. Marks are awarded for speaking and listening, reading, writing and spelling in the ratio $3:3:4:2$ respectively. Jasmine sat the test and scored 57%, 63%, 78% and 42% respectively for each part of the test. Find her average percentage for the English test.

4. Simon interviewed Annabel, Barbie and Cindy for the post of personal assistant. The criteria upon which he judged them, their relative importance and the scores they achieved are given in the table.

	Appearance	Personality	Telephone voice	ICT skills
Weighting	5	4	3	8
Annabel	4	8	9	6
Barbie	7	4	4	5
Cindy	3	3	8	8

    The person with the highest overall score was given the post; who was it?

5. A television production company needs a gofor (a person to go for something).
    The first impression an interviewee gives is from their clothes style. However it is four times more important that they are resourceful and three times more important that they have a good sense of humour.
    Each person interviewed is given a score out of ten for each of these criteria.

    Lawrence scored 9 for dress style, 7 for his resourcefulness and 5 for his sense of humour. Find his average score.

5.2   Index numbers

An index number shows the rate of change in price, quantity or value of an item or group of items over a period of time.

An index number is a percentage which gives the value or size of a quantity relative to a standard number, or base.

> **Hint**: The **standard number** is the value from a previous point in time known as the base year.

$$\text{Index number} = \frac{\text{quantity}}{\text{quantity in base year}} \times 100$$

Simple index numbers

A simple example of an index number is a **price relative**.

A price relative shows how the price of goods changes over time. It is calculated as a percentage of its value at a given base year.

$$\text{Price relative} = \frac{\text{price}}{\text{price in base year}} \times 100$$

Example

The table shows the price, to the nearest pound, of a toy as it rose and fell in popularity.

Year	1998	1999	2000
Price	8	15	6

Using 1998 as the base year, find the price relative in 1999 and 2000. Comment on these indexes.

The price relative for 1999 $= \dfrac{\text{price in 1999}}{\text{price in 1998}} \times 100$

$$= \frac{15 \times 100}{8} = 187.5$$

This means that the price of the toy in 1999 is 187.5% of its price in 1998. It is 87.5% more expensive.

The price relative for 2000 $= \dfrac{\text{price in 2000}}{\text{price in 1998}} \times 100$

$$= \frac{6 \times 100}{8} = 75 \text{ (the price relative)}$$

This means that the price of the toy in 2000 is 75% of its price in 1998. It is 25% ($= 100 - 75$) less expensive than it was in 1998.

The value of the index number (price relative) for the chosen base year is always given as 100.

In the example above this could be written as 1998 = 100.

● **Example**

A product's index numbers for three years are:

Year	1995	1996	1997
Index	100	105	126

(a) Which year is taken as the base year?
(b) What do the index numbers tell you about the price of the product?
(c) Find the price of the product in 1997 if its price in 1996 was £630.
(d) If 1996 is to become the new base year, find the new index number for 1997. Comment on this new index number.

(a) 1995 is the base year because its index is 100.
(b) The index numbers suggest that the price of the product is increasing year on year. There was a 5% increase from 1995 to 1996 and an increase of 26% from 1995 to 1997.
(c) In 1997 the price is 126% of the 1995 price: $126\% = £?$
   The price in 1996 is 105% of the 1995 price: $105\% = £630$

$$\frac{126}{105} = \frac{?}{630} \qquad \frac{126 \times 630}{105} = ? \qquad ? = £756$$

(d) If 1996 is the new base year then its index will be 100.

The index for 1997 relative to 1996 $= \dfrac{126}{105} \times 100 = 120$

This index number, 120, means that the price increased by 20% from 1996 to 1997.

Exercise 5B

1. The price of a 200 g jar of coffee in 1997 was £3. In 2000 the same jar of coffee cost £3.25. Calculate an index number (price relative) using 1997 as the base year.

2. In 1998 the price of a new small car (basic model) was £7400. The same car would have cost £6950 in 1996. Calculate a price relative using 1996 = 100.

3. In January 2000 unleaded petrol cost 84p/litre. In January 2001 it cost 77p/litre. Calculate a price relative using 2000 as the base year.

4. The price relative for a one-day travel card to London in 2001 is 116.7, using 1998 as the base year. If the price of the travel card in 1998 was £4.20, calculate its price, to the nearest 10p, in 2001.

5. The selling price of a scooter had an index of 132 in 1999 where 1995 = 100. If the selling price of the scooter in 1999 was £99, find its selling price in 1995.

6. A toyshop restocks each year after its summer sale and then always changes the price of its own brand toys. The year the shop opened is used as the base year. The table below gives the index numbers used by the shop for price changes.

Year	1992	1993	1994	1995	1996
Index	100	105	112	120	136

(a) In which year did the shop open?

(b) What do the index numbers tell you about the prices of the shop's own brand toys?

(c) If the shop's own brand tricycle cost £18 in 1992, how much would it cost in 1994?

(d) In 1996 the shop's own brand play-house cost £170. Find its selling price in 1994.

(e) If 1995 is to become the new base year, find the index number for 1996.

7. In 1993 the retail prices of a washing machine and a dishwasher were £180 and £300 respectively. In 1997 equivalent models cost £252 and £390 respectively.

(a) Using 1993 as the base year, calculate index numbers for each item.

(b) Which item had the greatest percentage increase in price?

(c) Which item had the greatest actual increase in price?

(d) Comment on your answers to parts (b) and (c).

5.3 Chain base numbers

To find out how the price of an item has changed over a year, you use the previous year as the base year.

> A chain base index number gives the relative value of an item using the previous year as the base year.
>
> A chain base index number tells you the annual percentage change.

Using chain base index numbers:

✦ The increases and decreases in price are always relative to the previous year.

✦ The base year changes for each calculation.

Example

The table shows the price changes over four years of a child's toy as it rose and fell in popularity:

Year	1998	1999	2000	2001
Price (£)	8	15	6	3

Find the chain base index number for each year.
What do the chain base index numbers show?

The index number for 1999 $= \frac{15}{8} \times 100 = 187.5$ (Base year is 1998)

The index number for 2000 $= \frac{6}{15} \times 100 = 40$ (Base year is 1999)

The index number for 2001 $= \frac{3}{6} \times 100 = 50$ (Base year is 2000)

This shows that the price of the child's toy increased by 87.5%, then decreased by 60% $(= 100 - 40)$, then decreased by 50% $(= 100 - 50)$.

Exercise 5C

1. The chain base index numbers for successive years for several different items are given below. Describe the price changes for each item during this time.

   (a) 150, 160, 170   (b) 98, 120, 156   (c) 100, 100, 100
   (d) 140, 100, 90   (e) 125, 80, 100

2. The prices of a child's shoes are given in the table.

Year	1990	1991	1992	1993	1994
Price (£)	12	20	25	29	35

   Calculate the annual percentage change (chain base index numbers) for the price of the shoes.

3. The list price of a second-hand car over successive years is as follows:

Year	1993	1994	1995	1996	1997
Price (£)	6000	3600	2500	1800	1450

   Use the chain base method to calculate index numbers for the price of the car. Comment on your results.
   Using 1993 as the base year, calculate an index number for 1996. Compare and comment upon your index numbers for 1996 using these two methods.

4. The index numbers for a toyshop's own brand toys are given in the table.

Year	1992	1993	1994	1995	1996
Index	100	105	112	120	136

Recalculate the index numbers using the chain base method to find the annual percentage change in the toy prices.

5. The index numbers in the table show the annual percentage change in price.

Year	1986	1987	1988	1989	1990
Index	100	120	124	124	110

Calculate the index numbers for 1987 to 1990 using 1986 as the base year. Comment on the 1990 price compared with the price in 1986.

Weighted index numbers

Bronze is made using 92% copper and 8% tin.
If the price of either element changes then the cost of producing bronze also changes. The change in cost is affected more by the cost of copper than by the cost of tin.

A price index for bronze needs to reflect the different proportions of copper and tin so you need a weighted index number to help calculate any change in cost.

To calculate a weighted index number you:

+ calculate the index number for each element, then
+ find the weighted average of those elements.

Weighted index number $= \dfrac{\Sigma \text{ (index number} \times \text{weight)}}{\Sigma \text{ (weights)}}$

Example

The table gives average price per ton for copper and tin for two years:

	1950	1955	Weight
Copper	£178	£351	92
Tin	£746	£740	8

Using 1950 as the base year, find the weighted average for the change in cost. What does the weighted average show?

Copper index number $= \dfrac{351}{178} \times 100 = 197.191\ldots$

Copper index number $\times$ weight $= 197.191\ldots \times 92 = 18\,141.57\ldots$

Tin index number $= \dfrac{740}{746} \times 100 = 99.1957\ldots$

Tin index number $\times$ weight $= 99.1957\ldots \times 8 = 793.5656\ldots$

Total weight $= \Sigma$ weights $= 92 + 8 = 100$.

$$\text{Weighted index number for bronze} = \frac{18\,141.57\ldots + 793.5656\ldots}{100}$$

$$= \frac{18\,935.1\ldots}{100}$$

$$= 189.35 \text{ (to 2 d.p.)}$$

This shows that the average price of the raw materials to produce bronze increased by 89.35% (to 2 d.p.) from 1950 to 1955.

> **Remember**: The price relative is the index number. You can see how to calculate it on page 145.

Exercise 5D

The table gives the market price in US dollars per tonne of six metals on the last trading day of the year. All prices have been rounded to the nearest $10.

	Copper	Nickel	Lead	Tin	Zinc
1992	2280	5920	450	5780	1060
1993	1770	5300	470	4760	1000
1994	3040	8870	650	6020	1140
1995	2800	7940	720	6280	1000
1996	2220	6350	700	5790	1050
1997	1720	5990	560	5400	1090

For questions 1–5, find the weighted index for the following alloys.

1. Brass (70% copper, 30% zinc) for 1994 with base year $= 1993$.

2. Bell metal (78% copper, 22% tin) for 1996 with base year $= 1995$.

3. Soft solder (60% tin, 40% lead) for 1995 with base year $= 1994$.

4. Gun metal (85% copper, 10% zinc, 5% tin) for 1993 with base year $= 1992$.

5. Statuary bronze (90% copper, 5% tin, 4% zinc, 1% lead) for 1996 with base year $= 1993$.

6. A company produces gilding metal (15% zinc, 85% copper) and Dutch metal (20% zinc, 80% copper). Calculate which of these two metal's costs changed the most from 1996 to 1997.

7. Calculate a chain base index for British 'silver' coins (75% copper, 25% nickel) from 1992 to 1997. Comment on the cost of producing British silver coins during this time.

8. £1 coins consist of 70% copper, 24.5% zinc and 5.5% nickel. Calculate a chain base index for £1 coins from 1992 to 1996.

Retail price index (RPI)

Weighted index numbers are not only used for alloys.

The **retail price index** is a form of weighted index which is used to monitor changes and make comparisons. However it only shows change in the cost of living of an 'average' person or family.

The retail price index is a weighted mean of the price relatives of goods and services.

Weightings are chosen to reflect the spending habits of an 'average' household.

A bakery would be interested in the weighted index for the ingredients of bread, pastries and so on for cost analysis.

In 1914 the RPI was based on what the government thought of as essential expenditure of working class families and assigned weights of food 60%, housing 16%, clothing 12%, heat & light 8% and miscellaneous 4%.

These weights became out of date with the rising standard of living. These categories needed revising with changes in lifestyle, for example entertainment, use of cars, holidays and so on.

Since the Second World War the index has been calculated on what households *actually* bought. In 1962 it was decided to revise the weightings annually, based on the actual expenditure of the previous three years.

Exercise 5E

Conduct a survey of expenditure for your class or a group of friends.
Decide how you will classify spending habits.
Find average prices for goods (magazines, CDs etc) bought now and over the past three years. (*Search the internet.*)
Calculate a retail price index for your group.
Discuss what implications this could have for pocket money increases, Saturday job pay rises and so on.

5.4   Population averages

The accurate recording and analysis of statistics, such as the number of births, deaths, marriages, crimes and accidents, are important for government departments to plan for the needs of housing, roads, education, health and so on.

These statistics are expressed as a number per thousand of the population. The actual total population is changing all the time as babies are born and people die, so the figure used can only be an estimate.

> Total population is the population at a specified time in the middle of the year.

> Population averages are given per 1000 of the population.

Crude rates

> $$\text{Crude (birth/death/}\dots\text{) rate} = \frac{\text{number of (births/deaths/}\dots\text{)}}{\text{total population}} \times 1000$$

● **Example**

There are 384 deaths in the town of Antville, which has a population of 16 000. Find the crude death rate.

$$\text{Crude death rate} = \frac{384}{16\,000} \times 1000 = 24$$

This means that there are 24 deaths per thousand of the population of Antville.

● **Example**

This table gives the population and number of deaths for Brighthove:

Age group	Population	Number of deaths
0–14	3000	27
15–44	4000	33
45–64	4000	42
65+	1000	45
Totals	12 000	147

Find the crude death rate for Brighthove.

$$\text{Crude death rate} = \frac{147 \times 1000}{12\,000} = 12.25$$

This means that there are 12.25 deaths per thousand of the population of Brighthove.

Exercise 5F

1. There were 4350 house crimes reported in Dodge City. If Dodge City had 20 000 households, calculate its crude house crime rate.

2. Last year in Dangerville, which had a population of 18 900, there were 2835 road traffic accidents. These resulted in 567 deaths.

   (a) Calculate the crude (road traffic) accident rate in Dangerville.
   (b) Calculate the crude (road traffic accident) death rate in Dangerville.

3. The crude birth rate in Cupston, which had a population of 8000 women of child-bearing age, was 62. How many births were there in Cupston?

4. The crude unemployment rate in a small village with a population of 2500 was 42.5. How many of the villagers were unemployed?

5. The table shows the number of births, age of mother and female population for a town.

Age (years)	Female population	Number of births
Under 18	16 000	322
18–44	45 000	1247
45 & over	34 000	31

   Using this information, calculate a crude birth rate for the town.

   Why is this crude birth rate a distortion of the true crude birth rate for the town?

Standardized rates

To make comparisons of population averages between different areas of the country, or between different countries, you need to consider the age distribution of each area.

> Standardized rates take account of the age distribution.

To calculate a standardized rate:

✦ calculate crude rates for each age group, then
✦ calculate a weighted mean average using 'standard population' data as the weights.

> Standardized rate $= \dfrac{\Sigma \,(\text{crude rate} \times \text{standard population})}{\Sigma \,\text{standard population}}$

The death rate in an area with a large number of elderly people is likely to be higher than in an area where the population is younger.

Standard population is the actual number or the percentage of the population for a given age group. It is found from data collected in a census.

Example

The crude death rates for Antville and Brighthove are 24 and 12.25. The crude death rate in Antville is almost double, making it appear a very unhealthy place to live.

Age group	Antville		Brighthove		Standard population
	Population	No. of deaths	Population	No. of deaths	
0–14	2000	30	3000	27	8
15–44	3200	36	4500	33	23
45–64	4800	102	3500	42	16
65+	6000	216	1000	45	13

> Standard population can be found in published resources such as census data.

Using the detailed information about the towns, find the standardized death rates for Antville and Brighthove. Comment on your answers.

First calculate the crude death rates for each age group in each town.

The crude death rates for each age group in Antville are:

$$\frac{30 \times 1000}{2000} \qquad \frac{36 \times 1000}{3200} \qquad \frac{102 \times 1000}{4800} \qquad \frac{216 \times 1000}{6000}$$

$$= 15 \qquad\qquad = 11.25 \qquad\quad = 21.25 \qquad\quad = 36$$

Then using the standard population as weights, calculate the standardized death rate:

$$\text{Standardized death rate} = \frac{(15 \times 8) + (11.25 \times 23) + (21.25 \times 16) + (36 \times 13)}{8 + 23 + 16 + 13}$$

$$= \frac{1186.75}{60} = 19.8$$

> The standardized death rate, 19.8, is lower than the crude death rate, 24, because a greater percentage of older people live in Antville than in the population generally.

The crude death rates for each age group in Brighthove are:

$$\frac{27 \times 1000}{3000} \qquad \frac{33 \times 1000}{4000} \qquad \frac{42 \times 1000}{3500} \qquad \frac{45 \times 1000}{1000}$$

$$= 9 \qquad\qquad = 8.25 \qquad\quad = 12 \qquad\qquad = 45$$

$$\text{Standardized death rate} = \frac{(9 \times 8) + (8.25 \times 23) + (12 \times 16) + (45 \times 13)}{8 + 23 + 16 + 13}$$

$$= \frac{1038.75}{60} = 17.3$$

> The standardized death rate for Brighthove is greater than the crude death rate (17.3 compared with 12.25) because of the smaller percentage of the oldest age group ($1000/12\,000 \times 100 = 8.33\%$) compared with the standard population ($\frac{13}{60} \times 100 = 21.67\%$) for that age group.

The standardized death rates for Antville and Brighthove are 19.8 and 17.3 per thousand respectively.
The standardized death rates for the towns are more similar than the crude death rates suggested.

Exercise 5G

1. The table shows the number of days absent at Wood Secondary School over one month.
   Calculate the standardized absence rate.

Year	No. days absent	Population	Standard population
7	47	156	25%
8	56	148	21%
9	102	132	18%
10	78	134	20%
11	52	120	16%

2. The table shows the number of people who developed skin cancer in Sun City and Cloudy Town.

	Sun City		Cloudy Town		
Age	Population	Skin cancer	Population	Skin cancer	Standard population
40 & under	52 000	480	3600	34	490
41–50	27 500	390	2800	56	160
51–60	11 000	890	2300	48	170
61–70	8000	820	1600	68	150
71 & over	1500	900	700	96	130

   Calculate standardized rates for developing skin cancer for Sun City and for Cloudy Town.
   Which place would you prefer to live? Give a reason for your choice.

3. The number of accidents for two building firms, Acmebuilt and Bodgitt, are given in the table.

	Acmebuilt		Bodgitt		
Age	Employees	Accidents	Employees	Accidents	Standard population
18–25	64	3	116	9	36%
26–35	70	5	64	7	26.8%
36–55	63	2	67	6	26%
56–65	28	2	28	2	11.2%

   Calculate standardized accident rates for both building firms.
   Which firm would you prefer to work for? Give a reason for your answer.

5.5   Time series

Time series data are observations of a variable over a period of time.

The variable could be temperature, sales, heating costs, number unemployed, …

The observations are usually made at regular time intervals.

Drawing graphs & seasonal variation

Time series data is displayed as a series of plotted points on a graph with time on the horizontal axis.

Example

The quarterly heating bills over three years for a household are given in the table.

	Jan.–March	April–June	July–Sept.	Oct.–Dec.
Quarter	1st	2nd	3rd	4th
1996	£80	£54	£47	£75
1997	£88	£68	£53	£82
1998	£97	No bill available	£60	£89

Plot the data on a time series graph.
What does your graph show?

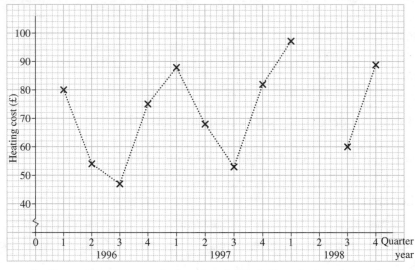

There are no heating bills issued to this household during the intervening months, so joining the points has no real meaning but is often done to help spot trends.

The graph shows that quarterly heating bills fall and rise throughout the year. This pattern is repeated the following year. This is due to seasonal variation and the need for more heating during the winter months.

> Seasonal variation is the regular rise and fall over a fixed period of time. It is sometimes called cyclical variation.

If you ignore the seasonal variations you can see the general trend of the data. For example, if you look at the first quarter only you will see that the general trend is that heating costs are rising.

> General trend is the underlying long-term trend. It is important in forecasting. It is sometimes called secular variation.

Predicting using trend lines

The general trend can be shown on the graph by drawing a trend line. Look at the general trend of the data and draw a line through the middle of the data.

Here is the trend line for the example on page 156.

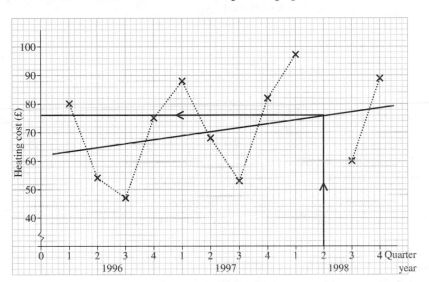

In this example the general trend is increasing. This means that your line should slope upwards. The data peaks and falls with every four points plotted so for each cycle there should be two points above the line and two points below the line.

This trend line can be used for prediction.

For example, you can predict that the second quarter of 1988 had a bill of around £76.

Sometimes a random event will cause an unusually large or small value. For example, the second quarter of 1998 may have been a very cold Spring. This would result in a different heating bill to the one predicted.

Chance events cause random variation.

Exercise 5H

1. The table shows the quarterly electricity bills for a bungalow over three years.

Quarter	1st	2nd	3rd	4th
1987	£70.00	£30.00	£20.00	£69.00
1988	£75.00	£34.50	£28.50	£75.00
1989	£79.00	£41.00	No bill available	£84.00

Plot the data on a time series graph and draw a trend line by eye. Use your trend line to estimate the missing electricity bill.

2. The table shows the half-yearly profits made by a toy shop.

Year	1996	1997	1998	1999	2000
Jan.–June	£21 000	£21 000	£24 500	£26 000	£27 500
July–Dec.	£30 000	£34 000	£35 000	£37 500	£39 000

Plot the data on a time series graph and draw a trend line by eye. Suggest a reason why the profits for the second half of the year should be more than in the first half of the year.

3. The numbers of people absent from work at a small factory were recorded.

	Monday	Tuesday	Wednesday	Thursday	Friday
Week 1	12	3	2	3	6
Week 2	11	3	3	4	7
Week 3	11	4	2	3	6

Plot the data on a time series graph and draw a trend line by eye. Suggest a reason why there should be more absenteeism on Monday and Friday.

4. A cinema manager had been asked to forecast how much profit he might make in the next month. He decided to do this by monitoring the number of people visiting his cinema over the past eight weeks.

Week	1	2	3	4	5	6	7	8
Attendance	9700	9400	10 500	9500	9600	9300	9550	9400

(a) Suggest a reason why attendance in week 3 is higher than the other weeks.

(b) Plot these values on a time series graph.
     Draw a trend line by eye. What does the trend line suggest?

(c) Discuss what other factors the cinema manager should consider when forecasting next month's profit.

Moving averages

An efficient and practical way of finding the trend and hence making accurate predictions is to find the moving averages for a set of data.

So long as you choose a whole cycle to average, this should eliminate any seasonal variations.

> Moving averages tend to reduce the amount of variation present in a time series, leaving only the trend.

You use a 4-point moving average when the data is based on four quarters in a year, for example phone bills, gas bills etc.

Example

(a) Calculate moving averages for the data in the example on page 156.

(b) Use these averages to draw a trend line for the data and use it to estimate:
(i) the bill for the second quarter of 1998,   (ii) the bill for the first quarter of 1999.

(a) Using a 4-point average, you calculate the average for every four consecutive data points.

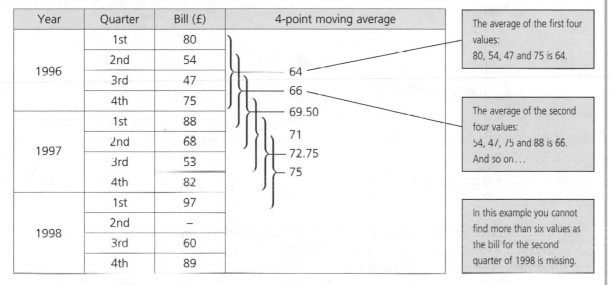

Year	Quarter	Bill (£)	4-point moving average
1996	1st	80	
	2nd	54	
			64
	3rd	47	
			66
	4th	75	
			69.50
1997	1st	88	
			71
	2nd	68	
			72.75
	3rd	53	
			75
	4th	82	
1998	1st	97	
	2nd	–	
	3rd	60	
	4th	89	

The average of the first four values:
80, 54, 47 and 75 is 64.

The average of the second four values:
54, 47, 75 and 88 is 66.
And so on...

In this example you cannot find more than six values as the bill for the second quarter of 1998 is missing.

Plot each of the averages at the midpoint of its range.

For example the first moving average, 64, is plotted midway between the second and third quarter of 1996. The second is plotted between the third and fourth quarter of 1996 and so on.

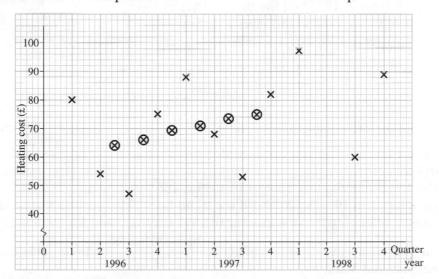

You need to plot the moving averages using a different colour from the time series data or circle them to tell easily which is which.

(b) Draw the best straight line you can through the moving averages. This is a trend line based upon moving averages.

> Note how much easier it is to draw the trend line using moving averages.

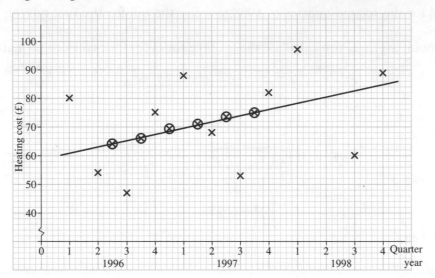

(i) To estimate the bill for the second quarter you need to consider where the moving average would have been and work out what the missing value is.
Let the missing value (second quarter 1998) be £y.

The moving average you would calculate would be:
$$\frac{53 + 82 + 97 + y}{4}$$

> 53, 82 and 97 are the bills for the third and fourth quarters in 1997 and the first and second quarters in 1998.

This moving average would have been plotted between the fourth quarter in 1997 and the first quarter in 1998. Reading from the trend line this gives £77. So:

$$\frac{53 + 82 + 97 + y}{4} = 77$$

$$\frac{232 + y}{4} = 77 \qquad 232 + y = 308 \qquad y = 76$$

An estimate for the missing bill is £76.

(ii) To predict a future value you can continue the trend line and extrapolate.
From the trend line the first quarter in 1999 is £88.
The first quarter in 1996 is £20 above the trend line. This is called **Seasonal Effect**.
The seasonal effect for the first quarter of 1997 is £18.50 and for 1998 is £18.

> A value from the graph **below** the trend line would be a **negative** number.

The first quarter average seasonal effect is $\frac{20 + 18.5 + 18}{3} = 18.83$. This is called Average Seasonal Variation.

The predicted estimated bill for the first quarter in 1999 is £88 + £18.83 = £106.83 (trend line value plus average seasonal variation).

Seasonal effect at a given data point is the difference between the actual value and the value read from the trend line at that point.

Average seasonal variation is the average of the seasonal effects for the same point in each cycle.

Exercise 5I —————————————————————————

1. (a) For the data given in Exercise 5H question 1, calculate appropriate moving averages.
   (b) Draw a new graph and plot the data given and the calculated moving averages.
   (c) Draw a trend line and use it to estimate the missing electricity bill.
   (d) Compare and comment on your answer with your previous answer.
   (e) Find the average seasonal variation for the first quarter.
   (f) Predict the value of the electricity bill for the first quarter of 1990.

2. (a) For the data given in Exercise 5H question 2, state why it would be appropriate to calculate 2-point moving averages.
   (b) Calculate 2-point moving averages for these data.
   (c) Draw a new graph and plot the original data and your moving averages.
   (d) Draw a trend line and use it to predict the half-yearly profits for 2001.

3. (a) For the data given in Exercise 5H question 3, calculate appropriate moving averages.
   (b) Draw a new graph and plot the original data and your calculated moving averages.
   (c) Use your moving averages to draw a trend line.
   (d) Use your trend line to predict the number of absentees on the Monday and Tuesday of the following week.

4. A street trader records his sales every four months. They were:

	1996	1997	1998	1999
Jan.–April	£315	£325	£330	£350
May–Aug.	£360	£375	£390	Figures missing
Sept.–Dec.	£410	£440	£465	£500

   (a) Plot the data on a time series graph.
   (b) Calculate appropriate moving averages and plot these on the same graph.
   (c) Use your moving averages to draw a trend line.
   (d) Use your trend line to estimate the missing sales figure for May–August 1999.
   (e) Find the average seasonal variation for January–April.
   (f) Predict the sales figure for the first four months of 2000.

5.6 Quality assurance

When items are manufactured there is usually some slight variability in the items produced.

To check that the variation is reasonable, firms use quality control techniques on samples taken from the production line. This is known as quality assurance or control.

To quality control items you:

✦ Take samples from the manufacturing process
✦ Find the average (usually the arithmetic mean) of the samples
✦ Plot the average against time
✦ Consider the variation from the graph.

Example

Size 10 trousers should have a waistband of 25 cm.

The manager of the Smoothpress factory collects the following data samples from size 10 trousers.

Time	1000	1100	1200	1300	1400	1500	1600
Mean waist of sample (cm)	25.1	24.9	24.9	25.2	25.0	24.8	25.2

Draw a graph and comment on the variability of the samples.

You draw the graph with the time axis starting at the mean waist size:

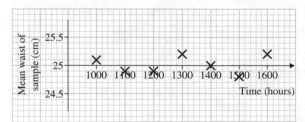

The graph shows that there is some variation. This is not excessive and so it may be assumed the manufacturing process is functioning normally.

When you consider the quality assurance, look out for:

The items are either all above or all below the required average.	The graph shows a general increase (or decrease).	There is an extreme value.	The control graph looks perfect.
		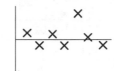	
This may mean the machine needs adjusting as it is producing items that are generally too small or too large.	The machine may need adjusting or the operator may need more training!	The machinery should be checked.	This may be too good to be true and another sample should be taken.

Exercise 5J

1. The manager of the Smoothpress factory carries out further quality control checks. He takes samples from five machines on the size 12 production line that should produce trousers with a waistband of 27 cm.

    For each machine's samples, draw a quality control graph and comment on the variability shown.

    What action would you advise him to take about the findings?

    (a) Machine A

Time	1000	1100	1200	1300	1400	1500	1600
Mean waist (cm)	27.1	27.0	27.2	27.0	27.1	27.2	26.8

    (b) Machine B

Time	1000	1100	1200	1300	1400	1500	1600
Mean waist (cm)	26.8	26.9	27.1	27.3	27.3	27.4	27.5

    (c) Machine C

Time	1000	1100	1200	1300	1400	1500	1600
Mean waist (cm)	27.1	27.0	27.0	26.9	27.2	27.6	27.9

    (d) Machine D

Time	1000	1100	1200	1300	1400	1500	1600
Mean waist (cm)	27.0	27.2	26.2	26.8	26.9	27.2	26.8

    (e) Machine E

Time	1000	1100	1200	1300	1400	1500	1600
Mean waist (cm)	27.0	27.0	27.1	26.9	27.0	26.9	27.0

2.   The median weight of samples taken every 30 minutes is recorded.
The target weight is 100 g.

Time	9.30	10.00	10.30	11.00	11.30	12.00	12.30
Mon	97	95	100	99	95	97	99
Tues	97	102	99	96	104	101	98
Wed	99	101	99	100	101	99	100
Thurs	110	105	103	102	98	98	93
Fri	102	98	101	100	91	93	94

For each of the five days state the time at which action, if any,
should be taken by the supervisor.
Explain why you believe action was necessary.

Summary

You should now be able to	Check out 5
1 Find the weighted average.	1 Find the weighted average of 72, 90 and 63 when weights are in the ratio $2:1:3$.
2 Calculate index numbers.	2 Calculate an index number for a drink that cost £2.50 in 2001 and £2 in 2000. 2000 is the base year.
3 Calculate crude and standardized population averages.	3 Calculate the crude death rate for a town population of 50 000 with 350 deaths.
4 Draw and interpret time series graphs.	4 Plot these data on a time series graph.

	1997	1998	1999	2000
Jan.–June	25	30	34	40
July–Dec.	60	66	75	80

Describe the trend shown by the graph.

You should now be able to	Check out 5
5 Find moving averages.	5 Find appropriate moving averages for these data and use them to draw a trend line.
6 Discuss quality assurance	6 A machine produces pencils 15 cm in length. The median of lengths of samples taken at hourly intervals are 13 cm, 14 cm, 14.5 cm, 13.5 cm, 19 cm, 15 cm. Draw a graph and comment on the variation shown.

Revision Exercise 5

1. A mathematics examination consists of three papers.
   The weighting of the papers is 3 : 2 : 1 as shown in the table.
   Two students, Sally and John, take the examination and their marks
   are shown in the table.

	Weighting	Sally	John
Paper A	3	60	40
Paper B	2	55	60
Paper C	1	40	?

   (a) Calculate the weighted mean mark for Sally.
   Sally and John had equal weighted mean marks.
   (b) Calculate the mark scored by John in Paper C.   [SEG]

2. Fabro plc uses four raw materials, A, B, C and D, in the
   manufacture of a product. The ratio, by weight, of the four raw
   materials, A, B, C and D, needed to produce each item is
   2 : 4 : 12 : 1, respectively. One kilogram of material D is used in the
   manufacture of one item. The costs of these raw materials in the
   years 1995–97 were as follows.

   Cost per kilo (£)

Raw material	1995	1996	1997
A	2.50	2.50	3.00
B	1.00	1.20	1.50
C	4.00	4.50	4.50
D	5.00	5.00	6.00

   (a) Show that the total raw material cost for **one** item in 1995 was
       £62.00.
   (b) Calculate the raw material cost of producing **one** equivalent
       item in 1997.
   (c) Use the results obtained in parts (a) and (b) to calculate, to
       one decimal place, a raw material cost index for 1997 using
       1995 as base.
   (d) An index of Fabro's labour costs is as follows.

Year	1995	1996	1997
Labour cost index	120	130	160

   Calculate the percentage increase in these costs from 1995 to 1997.

   (e) Compare the change in labour costs between 1995 and 1997
       with the change in raw material costs.   [NEAB]

3. The cost of a camera and a tin of cocoa are shown for the years 1992 and 1993.

	1992	1993
Camera	£145	£160
Cocoa	90p	120p

(a) Calculate the index number for the price of cocoa in 1993 using the 1992 base year index number as 100.

The 1994 index number for the camera price increase, using 1992 as the base year, was 120.

(b) Find the increase in price of the camera between 1993 and 1994. [SEG]

4. The table shows the indices for bicycle insurance costs. The base year for these indices is 1994.

Year	1994	1995	1996
Index	100	105	108

(a) What does the index number of 105 for 1995 tell you about insurance costs?

The cost of insuring a particular type of bicycle in 1995 was £31.50.

(b) (i) How much was the insurance in 1996?
(ii) What was the actual increase in cost for this insurance from 1994 to 1996?

In 1997 the insurance cost for this bicycle was £34.80.

(c) Calculate the index number for 1997. [SEG]

5. The price of a CD system was £480 in 1994 and £540 in 1995.

(a) Calculate the price index number for 1995 using 1994 as the base year.

The price index number for 1996 using 1994 as the base year was calculated to be 125.

(b) Calculate the price of the CD system in 1996.
(c) What was the percentage change in price of the CD system between 1994 and 1996?
(d) What was the percentage increase for the CD system during 1996? [SEG]

6. In January 1989 the cost per litre of petrol was 41.2 pence.
   In January 1994 the cost per litre had increased to 51.5 pence.

   (a) With 1989 as the base year, express the cost of petrol in
       January 1994 as an index number.
   (b) With 1994 as the base year, the index number for the cost of
       petrol in January 1996 was 120.
       Find the cost per litre of petrol in January 1996.
   (c) Explain the meaning of the index number 120.   [NEAB]

7. The table below shows some prices in 1972 and 1999. The index
   number compares 1999 prices with 1972 prices.

	1972	1999	Index number
Admission to a football ground	80p	£16	2000
Pint of beer	10p	£2.10	
Bottle of red wine	90p	£5.40	
Sliced white bread	6p	£0.60	

   (a) Show how the index number of 2000 in the table was obtained.
   (b) Copy the table and fill in the missing numbers.   [NEAB]

8. The table below shows the price of a mountain bike and a racing
   bike in 1988 and 1990.

	Price (£) 1988	Price (£) 1990	Price index (1990 relative to 1988)
Mountain bike	400	300	X
Racing bike	200	300	Y

   (a) (i) Find the value of X, the price index of a mountain bike.
       (ii) Find the value of Y, the price index of a racing bike.
   (b) In 1995 the price index (relative to 1988) of a mountain bike
       was 100. What can you say about the 1995 price of a mountain
       bike?   [NEAB]

9. The following table shows the percentage relatives and weight of
   certain commodities in 1995, taking 1993 as the base year.

   (a) Give **one** reason why a weighted average is sometimes more
       appropriate than the ordinary arithmetic average.
   (b) Given that the clothing bill in 1993 was £400, how much would
       it have been in 1995?
   (c) Use the information in the table to calculate a retail price
       index for 1995.

	Percentage relatives		Weight
	1993	1995	
Mortgage	100	110	0.4
Heat & lighting	100	130	0.2
Clothing	100	125	0.1
Food	100	115	0.25
Other items	100	120	0.05

[SEG]

10. The Managing Director of Crunch is keen to assess the impact that
retail prices may have on sales. He therefore obtains, in
summarized form, the following information from the Monthly
Digest of Statistics.

Item group	Single Item Index 1995	Single Item Index 1998	Weights
Food	250	291	208
Alcoholic drink	305	310	77
Tobacco	358	364	38
Housing	308	320	149
Fuel and light	391	402	67

(a) (i) Show that the total weighted index of retail prices for 1995 is 299.
(ii) Work out the equivalent weighted index of retail prices for
1998.
(iii) Hence calculate the all item (aggregate) weighted index of
retail prices for 1998 using 1995 as a base year.

(b) If the Food Group were excluded from the calculations in
(a) (iii), what effect would this have had on the resultant index?
Without further calculations justify your answer.   [NEAB]

11.

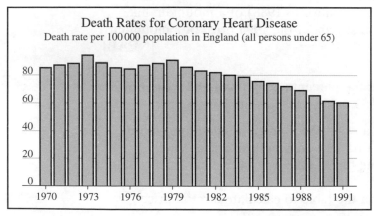

Source: *The Guardian*, 16 November 1993

The bar chart shows how the death rates for coronary heart disease have changed over the period 1970 to 1991.

(a) In which year was the highest death rate recorded?

The death rate for 1983 was 80.

In 1983 Burntwood, with a population of 85 000, recorded 71 deaths from heart disease.

(b) By suitable calculations show whether the 71 deaths recorded is higher or lower than the number you would expect from the Burntwood area.

The diagram shows a downward trend in death rates.

(c) Indicate a likely reason for this downward trend.   [SEG]

12. The population of Alphaville was 7500 at the beginning of 1996.

(a) The crude birth rate per thousand for 1996 was exactly 3.2. How many births were there in 1996?

(b) The crude death rate per thousand (correct to 1 decimal place) for 1996 was 2.5. How many deaths were there in 1996?

(c) What was the population of Alphaville at the end of 1996? (Assume that no one left or moved into the town.)   [NEAB]

13. The population of Gammaville was 17 500 at the beginning of 1998.

(a) The crude birth rate per thousand for 1998 was exactly 3.6. How many births were there in 1998?

(b) The crude death rate per thousand (correct to 2 decimal places) for 1998 was 2.74. How many deaths were there in 1998?   [NEAB]

14. The data in the following table are for the town of Westport, for the year 1992.

Age group	% Population of Westport	Death rate per 1000	Standard population %
0–	12.0	2	15.0
10–	12.5	1	15.5
20–	23.5	3	27.5
40–	25.5	14	23.5
60–	25.0	74	17.5
80+	1.5	244	1.0

The total population of Westport at the beginning of 1992 was 25 000 and 375 people died during 1992.

(a) Calculate the crude death rate per 1000.

The crude death rate for the whole of the UK for 1992 was 12.5.

(b) Give a possible reason for the difference between the crude
     death rates of Westport and the UK as a whole.
(c) Calculate the standardized death rate for Westport for 1992.
(d) Give a reason why you would not expect your answers to (a)
     and (c) to be the same.   [SEG]

15. The table shows the age distribution of a town X, the number of
     deaths for each age group and the percentage of each age group in
     the whole country.

Age group	Population in 1000s	Number of deaths	Standard population
Under 10	20	324	15%
10–24	32	125	25%
25–44	40	268	25%
45–64	23	426	20%
65 and over	12	560	15%

(a) Calculate the crude death rate for town X.

The crude death rate for a town Y is 19.2 per thousand.

(b) How would you expect the age distribution of town Y to differ
     from town X?
(c) Calculate the standardized death rate for town X.
(d) Explain why the standardized death rate is a better
     representation that the crude death rate.   [SEG]

16. A local newspaper suggested that Westhope is a healthier place to
     live than Martrent.
     Data for the two towns is given below.

| Age group | Westhope | | Martrent | | Standard population |
	No. of deaths	Population	Death rate per 1000	Percentage of population	Percentage of population
0–	6	1500	8	40%	25%
20–	4	2000	2	25%	30%
40–	24	1600	10	30%	30%
60–	100	2000	90	5%	15%

(a) Find the crude death rates per thousand for each group in Westhope.
(b) Calculate the standardized death rates for the two towns.
(c) With reference to your answer in (b), is the newspaper correct in its suggestion that Westhope is a healthier place to live than Martrent? Give a reason for your answer.   [NEAB]

17. The table shows the number of cars produced in Britain, in the years shown, from 1984 to 1996.

Year	1984	1986	1988	1990	1992	1994	1996
Production of cars (thousands)	900	1000	1250	1290	1290	1500	1690

(a) Plot this information on a graph.
(b) Draw a trend line on your graph.
(c) Use your graph to estimate
    (i)   the number of cars produced in 1985,
    (ii) the number of cars that will be produced in 1998. [NEAB]

18. The table below shows the number of licences issued for televisions from 1986 to 1993.

Year (X)	1986	1987	1988	1989	1990	1991	1992	1993
Number in millions (Y)	19.6	20.3	21.2	21.6	22.0	22.4	22.8	23.1

(a) Plot the time series on a graph.
(b) Draw the trend line by eye.
(c) Use your trend line to estimate the number of licences issued in 1994.
(d) State, with a reason, whether your answer to part (c) is likely to be too high or too low.   [NEAB]

19. A school canteen manager notes the quarterly turnover as follows.

	Jan–Mar	Apr–June	July–Sept	Oct–Dec
1992	£7500	£5500	£2000	£8200
1993	£6700	£4300	£1600	£8200
1994	£5100	£3900	£1200	£7500

(a) Which year has the biggest turnover?
(b) Give a reason why the third quarter turnover is the lowest in each year.
(c) Plot the data on a graph.
(d) Draw the trend line by eye.
(e) What does the trend line tell you about the canteen's turnover?
          [SEG]

20. Leena wanted to estimate how much her next electricity bill might be.
She found her last 11 bills.
The table shows, in £, her last 11 quarterly bills.

	Spring	Summer	Autumn	Winter
1992	95	60	110	155
1993	108	66	120	170
1994	110	80	135	

(a) Plot these data on a graph.
(b) Calculate the 4-point moving average for these data.
(c) Plot the moving averages on the graph and draw the trend line.
(d) Use your trend line to predict the bill for Winter 1994.   [SEG]

21. The number of ties sold in a shop, in each four-month period, is
shown.

Year	Period	Number of ties sold
	1	44
1991	2	120
	3	68
	1	36
1992	2	196
	3	48
	1	48
1993	2	116
	3	76
	1	32
1994	2	136
	3	120

(a) Plot these data on a time series graph.

Moving averages are to be calculated for these data.

(b) Explain why a 3-point moving average must be used.
(c) Calculate the 3-point moving averages.
(d) Plot your moving averages on your graph and draw a line of
best fit.
(e) What is the statistical term used to describe the line of best fit?
[SEG]

22. The following data give the quarterly sales, in £10 000's, of gardening equipment at the Green Fingers Garden Centre over a period of four years.

	Quarter			
	1st	2nd	3rd	4th
1992	20	26	24	18
1993	24	30	27	23
1994	26	34	31	25
1995	30	36	35	29

(a) Plot these values on a graph.
(b) Suggest a reason for the seasonal variation shown by your graph.
(c) Calculate the four-point moving averages for these data.
(d) Plot these moving averages on your graph.
(e) On your graph, draw a trend line by eye.
(f) Use your graph to estimate the sales during the first quarter of 1996.   [SEG]

23. The table shows the amounts, in £1000s, deposited in a bank on each of 12 weekdays before Christmas.

	Mon	Tue	Wed	Thur	Fri
Week beginning 6/12	250	190	200	215	230
Week beginning 13/12	280	215	235	240	255
Week beginning 20/12	305	245			

(a) On a graph, plot the daily amount deposited.
(b) Calculate the 5-point moving averages for the data.
(c) Explain why a 5-point moving average is appropriate.
(d) Plot the moving averages on the graph and draw the trend line.
(e) Calculate the average daily variation for Wednesday.
(f) Use the daily variation in order to estimate the sum deposited on the next Wednesday.   [SEG]

24. The table below shows the number of units of electricity used by a householder during eight successive quarters in 1995 and 1996.

Year	Electricity consumption (units)			
	Quarter 1	Quarter 2	Quarter 3	Quarter 4
1995	1450	1080	730	1280
1996	1630	1220	930	1460

(a) Draw a time-series graph to represent the amounts of electricity used over the eight quarters.

(b) (i) What are the main seasonal trends shown by your graph?
    (ii) What is the most likely explanation for them?

(c) Calculate appropriate moving averages for the data.

(d) Plot the moving averages on your graph.

(e) Draw by eye the trend line on the graph.

(f) The seasonal effects for Quarters 1 and 2 are as follows:

Seasonal effect	
Quarter 1	Quarter 2
+390	−77.5

Use this information and the trend line to provide seasonally adjusted forecasts for Quarter 1 and Quarter 2 of 1997.

6 Correlation and regression

> *First get your facts;*
> *then you can distort them at your leisure.*
> Mark Twain

The bigger the car, the more petrol it uses but the fewer miles per gallon it will do:

This unit will show you how to

- ✦ Draw scatter diagrams and trend lines
- ✦ Interpret scatter diagrams
- ✦ Interpolate and extrapolate data from graphs
- ✦ Calculate Spearman's coefficient of rank correlation

Before you start

You need to know how to	Check in 6
1 How to draw a graph	**1** Draw a graph with x-values from 0 to 6 and y-values from 0 to 5. Plot and join the points (2, 1), (6, 1), (4, 5). What shape is this?
2 How to find the mean (See Unit 4 page 91)	**2** Find the mean of 17.1, 34.8, 27.1, 8.9 and 22.5.
3 How to calculate the equation of a line. The equation of a line with gradient 2 and y-intercept 3 is $y = 2x + 3$.	**3** Find the equations of the lines with (a) gradient 5, y-intercept 4 (b) gradient −3, y-intercept 1 (c) gradient 3, y-intercept −8 (d) gradient −5, y-intercept −4

6.1 Scatter diagrams

A **scatter diagram** shows the relationship between two variables. It is sometimes called a scatter graph or scattergram.

To draw a scatter diagram you plot points on a graph.

Remember: Variables are quantities that vary.

Example

Martin made and sold ice-cream and he wanted to predict how much he needed to make each day.

He believes he sells more when the weather is hotter.

He recorded the maximum temperature and the ice-cream sales every day for eight days.

His results are summarized in the table.

Temperature (°C)	16	15	18	14	21	25	23	24
Sales in £s	75	65	80	60	100	145	130	135

Draw a scatter diagram of this data.
Do you agree with Martin's claim?

Plot the points: (16, 75) (15, 65) (18, 80) (14, 60) (21, 100) (25, 145) (23, 130) (24, 135)

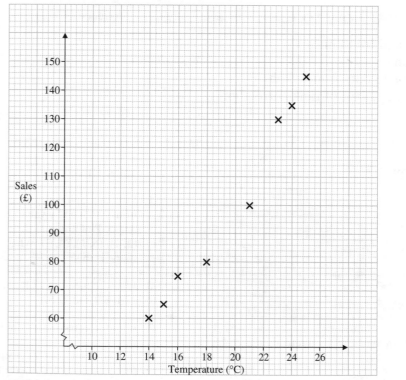

Martin can use the graph to predict sales depending on the temperature.
You can see how to do this in section 6.3.

The graph shows that the sales increase as the temperature increases so, yes, Martin's claim seems correct.

Recognizing correlation

You use a scatter diagram, or scattergraph, to investigate if there is a link, or **correlation**, between two variables.

On a scatter diagram:

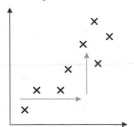

If one variable increases as the other variable increases then there is **positive correlation**.

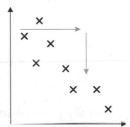

If one variable decreases as the other variable increases then there is **negative correlation**.

Martin's data showed positive correlation.

There is no correlation when the points are scattered showing no pattern.

For example, Martin asked the age of every child who bought ice-cream.

He plotted his data on a scattergraph:

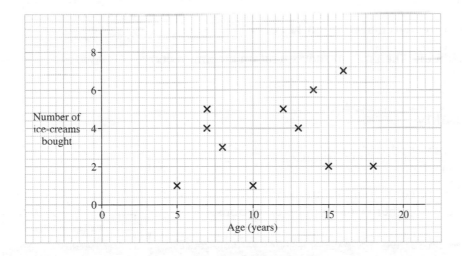

The points are scattered with no pattern. This shows there is no link between the age of a person and the number of ice-creams they buy. There is no correlation.

On a scatter diagram, if the plotted points do not show any pattern then there is no correlation between the variables.

Exercise 6A

For questions 1–3 choose the most appropriate scatter diagram for the situation: A, B or C.

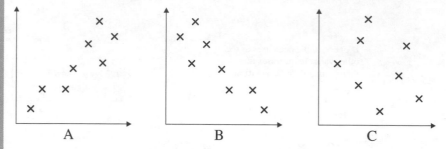

1. As you get older your reaction speed slows down.

2. People who are good at maths are usually good at music.

3. There is no connection between height and hair length of 14-year-old girls.

For questions 4–6, draw a sketch of a suitable scatter diagram, labelling the axes with the appropriate variables.

4. The outside temperature and the amount of gas used for central heating for a house.

5. The weight and age for children aged up to 12 years.

6. The height of a person and the amount they spend on cinema visits each year.

Correlation and causality

> **Correlation** implies a connection between two variables.

For example, people who are good at maths are usually good at music. This is a general trend to which there will often be exceptions.

> **Causality** implies a direct link between two variables. One variable **causes** the change in the other variable.

Scientific experiments often show a causal link between variables.

For example, 'the outside temperature and the amount of gas used for central heating'.

The lower the temperature, the greater the amount of gas used: one variable directly causes the other to change.

Sometimes there is no direct link between two variables, but they are connected by a third variable.

For example, in the last twenty years the number of microwave ovens and the number of television sets have both increased.

One is not directly related to the other, but they are both related to changes in technology.

You can sometimes find correlation when there is no real link. This is known as spurious correlation.

This is a true story...

Data was collected about the number of babies born in different areas and the number of storks nesting on houses in those areas.

When the data was plotted, the scatter diagram showed positive correlation.

But you cannot assume that more storks nesting mean more babies!

> It is possible to find correlation between variables that are unlikely to be connected. This is spurious correlation.

> This is more likely if you only collect a small amount of data.

Exercise 6B

For the following pairs of variables, decide whether they are likely to be linked by:

✦ correlation
✦ causality
✦ a third variable

1. The lengths of a sample of babies and their head circumference.

2. The load attached to a spring and its extended length.

3. The number of deckchairs on a particular beach and the number of ice-cream sales.

4. The height of a sample of students and their shoe size.

5. The volume of traffic and the number of road accidents.

6.2 Lines of best fit

If a scatter diagram seems to show correlation you can often draw a straight line through the points. This is called a line of best fit.

> You can draw a line of best fit on a scatter diagram if the graph shows that there is correlation.

> The line of best fit will pass through the mean average of each data set.

You need to draw a line that best fits the data.

Example

Callum wanted to know how high his power ball would bounce.

He dropped the ball with force from various heights and measured the bounce height.

His results are summarized in the table.

Drop height (cm)	30	40	50	60	70	80	90	100
Bounce height (cm)	27	48	53	62	69	89	102	110

(a) Draw a scatter diagram of this data.
(b) Describe the correlation and draw a line of best fit.

Plot these points: (30, 27) (40, 48) (50, 53) (60, 62) (70, 69) (80, 89) (90, 102) (100, 110)

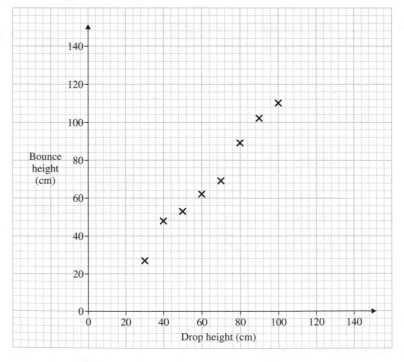

Hint: In an experiment the variable you **control** always goes on the *x*-axis.
The variable you measure as a **result** always goes on the *y*-axis.

(b) You can see that as the height from which Callum dropped the power ball increased, the bounce height also increased. This means that there is positive correlation between the drop height and the bounce height.

As the points appear to lie close to a straight line, you can draw a line of best fit.

First find the mean of each variable:

Mean drop height $= \dfrac{30 + 40 + 50 + 60 + 70 + 80 + 90 + 100}{8}$

$= \dfrac{520}{8} = 65\,\text{cm}$

Mean bounce height $= \dfrac{27 + 48 + 53 + 62 + 69 + 89 + 102 + 110}{8}$

$= \dfrac{560}{8} = 70\,\text{cm}$

> Remember: the mean is explained on page 91.

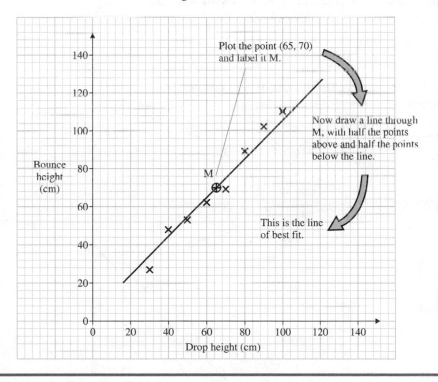

Plot the point (65, 70) and label it M.

Now draw a line through M, with half the points above and half the points below the line.

This is the line of best fit.

Exercise 6C

For each of the following questions draw a scatter graph to represent the data.

If appropriate, draw the line of best fit on your graph. Remember to find the mean value first.

1. Carrie conducted an experiment to find out if height and arm span were related. She measured the heights and arm spans in centimetres of eight of her friends.

Height (cm)	136	158	126	131	149	157	143	152
Arm span (cm)	134	146	122	130	147	150	134	145

2. Theo was given a small collection of old pennies. He noted down how old each penny was and how much it weighed.

Age (years)	51	47	54	33	39	46	42	49	36
Weight (grams)	7.2	9.8	6.1	12.0	10.2	8.5	9.7	7.4	11.6

3. Ashish compared the published price of several children's books with the price of the same books in a discount book club catalogue.

Published price	£10.00	£6.00	£11.00	£15.00	£13.00	£5.00	£10.00
Book club price	£7.00	£4.50	£8.50	£11.00	£9.00	£4.00	£8.00

4. The following table summarizes the number of staff absent and the number of thefts each day for two working weeks from a department store.

Absentees	32	19	25	29	11	14	23	12	30	24	16	2
Thefts	16	18	27	8	21	22	5	28	21	17	17	9

The assistant manager has to investigate if there is any connection between the two variables.

Strong or moderate correlation?

In Callum's experiment his results lie on an almost perfectly straight line.

This is **strong positive correlation**.

> If almost all plotted points lie very close to a line of best fit, then correlation is strong.

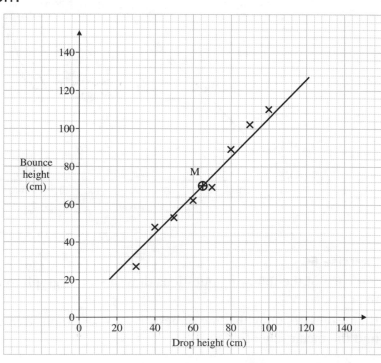

If plotted points tend to lie in a line, but not very close to that line, then correlation is moderate.

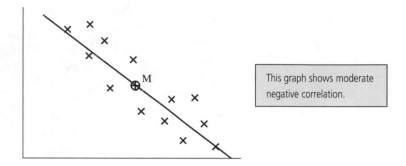

This graph shows moderate negative correlation.

Example

Barry carried out an experiment to find the reaction time of 30 volunteers after they had consumed alcohol. His results are summarized in the scatter graph.

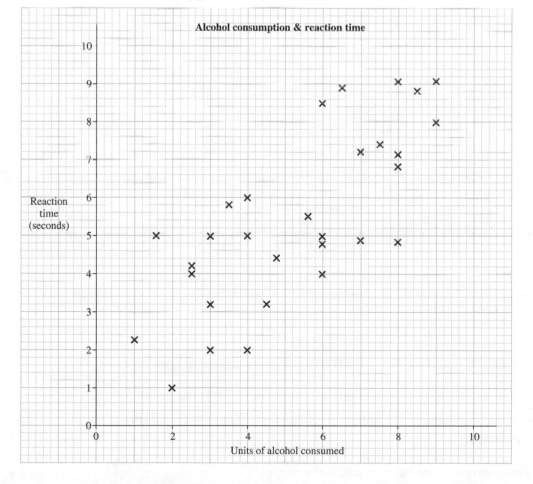

Describe the correlation suggested by the data.

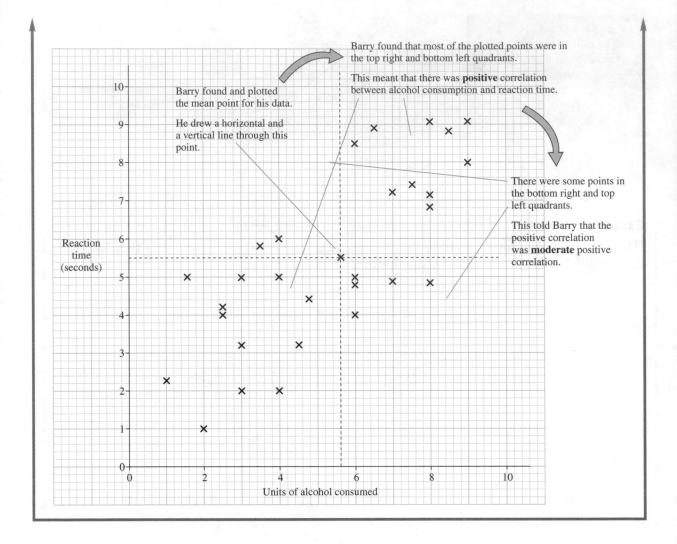

Exercise 6D _____

For your graphs in Exercise 6C questions 1 to 4, describe the correlation shown.

6.3 Interpolation and extrapolation ▬▬▬

You can use a line of best fit to estimate other values from the graph.

Finding estimates from a line of best fit is known as either **interpolation** or **extrapolation**.

> Interpolation is an estimate from **within** the range of given *x*-values.

> Extrapolation is an estimate from **outside** the range of given *x*-values.

Example

Tomris thought that there might be a connection between the life expectancy of mammals and their gestation period.

The scatter graph shows the results of her investigation.

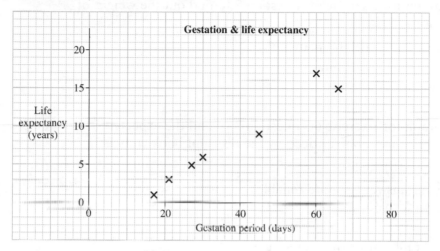

(a) Draw a line of best fit on the graph.

(b) Use the line to estimate the life expectancy of a mammal with a gestation period of:
   (i) 50 days,   (ii) 75 days.

Note: The graph shows moderate positive correlation so a line of best fit can be drawn.

(b) (i) 12 years,
is an estimate of the life expectancy of a mammal with a gestation period of 50 days.

50 days lies within the range of data so this is interpolation.

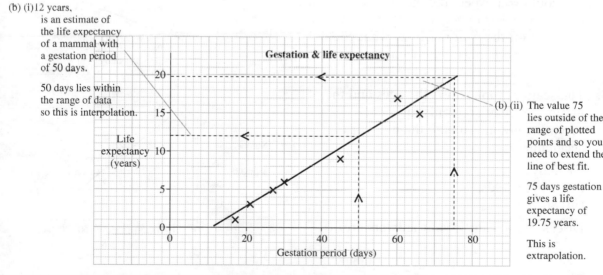

(b) (ii) The value 75 lies outside of the range of plotted points and so you need to extend the line of best fit.

75 days gestation gives a life expectancy of 19.75 years.

This is extrapolation.

You need to be careful when extrapolating data from a graph because it does not always make sense to extend a line of best fit.

When extrapolating data always question if the answer is realistic.
The further you extrapolate, the less reliable your estimate.

Exercise 6E _____

1. Use your graph from Exercise 6C question 1 to find out the arm span of another of Carrie's friends who is 146 cm tall.

2. Use your graph for Exercise 6C question 2 to find out the weight of a penny that is:

   (a) 44 years old,   (b) 60 years old.

   Which of your answers is more reliable?
   Give a reason for your answer.

3. Use your graph for Exercise 6C question 3 to find the expected publisher's price for a book priced by the discount book club as:

   (a) £15.00,   (b) £6.50.

   Which of your answers is more reliable?
   Give a reason for your answer.

4. The scatter graph shows the monies gambled on a fruit machine and the amount paid out to eleven different people.

   (a) How much did the unluckiest player lose?
   (b) If you decide to gamble £11, how much could you expect the machine to pay out?
   (c) Tabitha has only got £5. If she gambles it all on the fruit machine, what could she expect to win? Explain your answer.

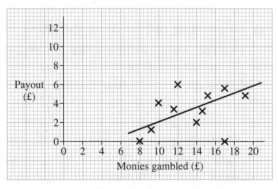

5. The scatter graph shows the ages and the number of road accidents for men.
   From the graph estimate the number of accidents for

   (i)   35-year-olds,
   (ii) 12-year-olds,
   (iii) 80-year-olds.

   Discuss how realistic your answers are.

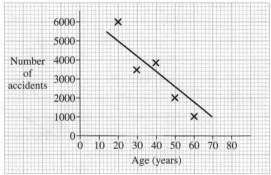

6. The graph shows the length of pregnancy, in days, and the weights of premature babies, in kg.
   Why would it not be sensible to use this graph to estimate the weight of a baby whose mother had been pregnant for 320 days?

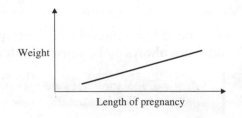

6.4   The equation of a line of best fit

Finding the equation of a line of best fit is exactly the same as finding the equation of any straight line.

You can calculate the equation of a line by equating gradients at two places.

You could also find the equation by finding:

✦   the gradient
✦   where the line crosses the y-axis, the y-intercept.

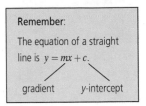

Remember:

The equation of a straight line is $y = mx + c$.

gradient y-intercept

● Example

This is Callum's graph from page 181.

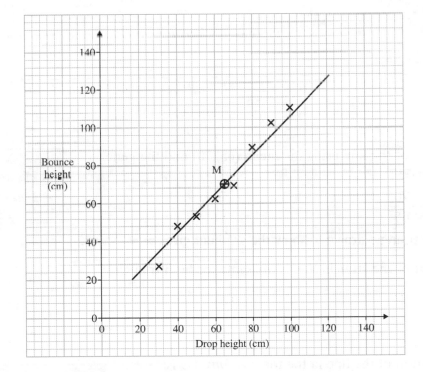

(a) Calculate the equation of the line of best fit.

(b) Use this equation to predict the bounce height (y) for a drop height (x) of 83 cm.

(c) What does the gradient mean in the context of the question?

(d) What does the y-intercept mean in the context of the question? Is it sensible to extend the graph to $x = 0$?

(a) To find the gradient, triangles are constructed on the line of best fit at two different places.

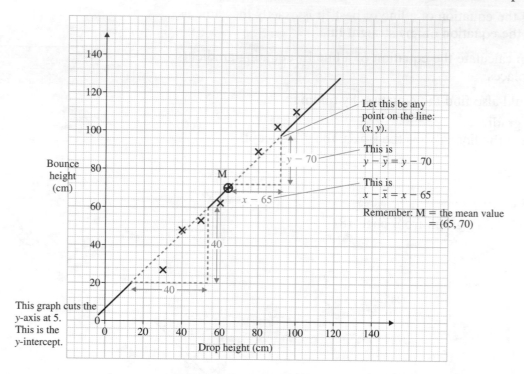

The gradient of the line using the mean (65, 70) and the general point (x, y) is

$$m = \frac{y - 70}{x - 65}$$

> The gradient of a straight line is the vertical difference divided by the horizontal difference.

The second triangle drawn to show the gradient gives $m = \frac{40}{40} = 1$.

The gradient of the straight line is the same for both triangles:

$$\frac{y - 70}{x - 65} = 1$$

$$y - 70 = x - 65$$

So the equation of the line in the form $y = mx + c$ is:

$$y = x + 5$$

The gradient is 1   The y-intercept is 5

(b) When $x = 83$, $y = 83 + 5$

$$= 88$$

So the predicted bounce height is 88 cm.

> Remember: This is an interpolated prediction since it is within the range of data.

(c) The gradient is 1.
This tells you that an increase in drop height of 1 cm will lead to an increase in bounce height of 1 cm.

(d) The y-intercept is 5.
This tells you that a drop height of zero will give a bounce height of 5 cm. This is clearly unrealistic, so the line should not be extended to $x = 0$.

> Remember: $x = 0$ is an extrapolated value outside the range of data.

Exercise 6F

1. The points (1, 7) and (3, 15) lie on a line of best fit.
Calculate the equation of the line.

2. Find the equation of the line that passes through the points (1, 5) and (3, 1).

3. A line of best fit passes through the points (0.8, 4.4) and (2, 2.6).
Calculate the equation of the line.

4. A line of best fit passes through the points (4.8, 10.17) and (3.4, 6.11).
Calculate the equation of the line.

5. (a) Calculate the equation of your lines of best fit for:
   (i)   Exercise 6C question 1
   (ii) Exercise 6C question 2
   (iii) Exercise 6C question 3.

   (b) For each line
   (i) Explain what the gradient means in the context of the question.
   (ii) Comment on whether the y-intercept gives a realistic value.

Non-linear data

You cannot always draw a straight line through plotted points on a scatter graph.

Sometimes the points lie on a curve, and you say that the relationship is **non-linear**.

> For straight-line graphs the relationship between the variables is **linear**.

Equations of typical curves include:

$$y = \frac{a}{x} + b, \quad y = ax^2 + b \quad \text{or} \quad y = a\sqrt{x} + b \quad \text{where } a \text{ and } b \text{ are constants.}$$

Their graphs look like this:

$y = \dfrac{a}{x} + b$

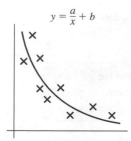

$y = ax^2 + b$

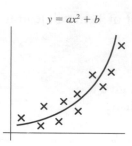

$y = a\sqrt{x} + b$

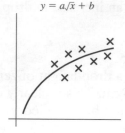

You can change all these curves into straight lines by plotting appropriate graphs. This will help you to find the values of a and b, and then you can find an equation for the graph.

> Changing non-linear data to straight lines is called reducing to linear form.

If the data resembles $y = \dfrac{a}{x} + b$, plot y against $\dfrac{1}{x}$

If the data resembles $y = ax^2 + b$, plot y against x^2

If the data resembles $y = a\sqrt{x} + b$, plot y against $\sqrt{x}$

The gradient of each graph will be a and the y-intercept will be b.

Example

The heating cost (C) of hotels, varies with the number of rooms (N).

Number, N	10	20	50	75	100
Cost, C	245	860	5070	11 300	19 970

The scatter graph looks like a curve of the type $y = ax^2 + b$.

> If you are asked to reduce non-linear data to a linear form, a suggested relationship will be given.

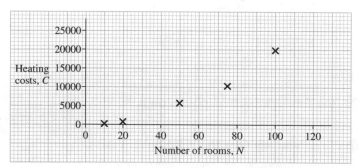

Reduce the data to linear form and draw a new scatter graph.

In this graph C is on the y-axis and N on the x-axis, so $C = aN^2 + b$.

To make this data linear, you need to plot C against N^2 (instead of C against N).

N	10	20	50	75	100
N^2	100	400	2500	5625	10 000
C	245	860	5070	11 300	19 970

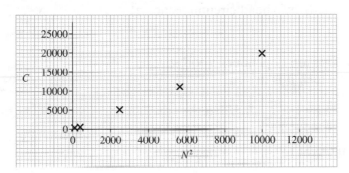

The heating cost graph using N^2 in place of N looks linear.

The data has been reduced to a linear form.

Now follow the method in the example on page 188 to help you find the values of a and b.

Then you can use the equation $C = aN^2 + b$ to make predictions.

Exercise 6G

1. The temperature of a hot drink is measured at one-minute intervals.

Time, s minutes	1	2	3	4	5	6	7	8	9	10
Temperature, $T\,°C$	89	46	30	25	20	16	13	11	10	9

It is thought that the temperature, T, of the drink is connected to time, s, by the formula $T = \dfrac{a}{s}$ where a is a constant.

Draw a scatter graph of T against $\dfrac{1}{s}$ to check this assumption.

Use your graph to find the value of a.

Using your graph or equation, estimate the temperature of the drink after:

(a) 1.5 minutes,   (b) 4.5 minutes,   (c) 15 minutes.

2. The head circumference of a baby is believed to be governed by the formula $C = \sqrt{A}$ where A is the baby's age in months. The following data was collected by a mother.

Age A (months)	1	2	4	6	8	9
Head circumference C (cm)	36.4	39.3	42	44	45.4	46

(a) Draw a scatter diagram of C against $\sqrt{A}$ to verify the formula.
(b) Use your graph to predict the baby's head circumference at 12 months.

3. The running costs, C, of a minibus are dependent upon several overheads and the speed, S, at which it is driven.

Speed, S	28	30	35	40	50	55
Running costs, C	20	21	24	30	35	42

The running cost equation is thought to be $C = aS^2 + b$, where a and b are constants.

Draw a scatter graph to verify this equation.
Use your graph to find the values of a and b.

6.5 Spearman's rank correlation coefficient

Spearman's rank correlation coefficient, r_s, is a measure of the agreement between two data sets.

It is used to find the extent to which two sets of data correlate.

r_s is measured on a scale from $^-1$ to $^+1$.

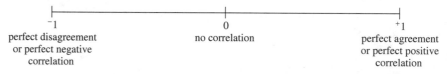

$^-1$
perfect disagreement
or perfect negative
correlation

0
no correlation

$^+1$
perfect agreement
or perfect positive
correlation

Each data value is assigned a **rank** depending on its size within its data set.

r_s is based on the **differences**, d, between corresponding ranks.

Spearman's rank correlation coefficient, $r_s = 1 - \dfrac{6\Sigma d^2}{n(n^2 - 1)}$

where d is the difference between corresponding ranks
and n is the number of pairs of data.

Remember:
Σ means 'sum of'.

The closer r_s is to $^+1$, the more agreement there is between the ranks.
The closer r_s is to $^-1$, the more disagreement there is.
If r_s is close to zero, the ranks neither agree nor disagree.

Example

Contestants in a gymnastics competition are awarded marks by several judges.

Two of the judges are chosen to see how well their marks correlate.

Their marks are:

Contestant	Abi	Betty	Carly	Debs	Eileen	Fran	Gina	Hannah
Judge Ment	4.5	5.6	3.9	5.2	4.6	5.5	5.9	5.1
Judge Ship	5.3	5.9	4.5	5.6	4.9	6.2	6.7	6.3

Both judges agree the best and worst placed contestant, but you can see that in general Judge Ship is more generous with his marks.

Calculate Spearman's rank correlation coefficient for this data.

✦ First rank the marks in their place order:

Contestant	Abi	Betty	Carly	Debs	Eileen	Fran	Gina	Hannah
Judge Ment	7	2	8	4	6	3	1	5
Judge Ship	6	4	8	5	7	3	1	2

✦ Subtract corresponding ranks to get the differences d:

d	1	2	0	1	1	0	0	3

Note: You only need to find the actual difference, not whether it is positive or negative.

Square the differences, d^2

d^2	1	4	0	1	1	0	0	9

$\Sigma d^2 = 16$

There are 8 contestants, so $n = 8$.

$$r_s = 1 - \frac{6\Sigma d^2}{n(n^2 - 1)}$$

$$= 1 - \frac{6 \times 16}{8(64 - 1)}$$

$$= 1 - \frac{96}{504}$$

$$= 1 - 0.1905 = 0.8095 \text{ or } 0.81 \text{ to } 2\,\text{d.p.}$$

0.81 is close to 1 so there is a fairly high agreement between the two judges.

Note: r only measures the agreement between **ranks**; it does not give you the correlation between the actual values themselves.

If two data values within a set are equal you can use the method for **tied ranks** outlined in the following example.

Example

Twins Kevin and Perry, whose tastes usually agree, were asked to taste test 10 drinks. They independently awarded marks out of 40 for the drinks.

Drink	A	B	C	D	E	F	G	H	I	J
Kevin	36	10	25	8	12	29	25	33	22	21
Perry	34	27	31	23	13	33	27	27	27	11

Find Spearman's rank correlation coefficient for this data.

For Kevin, C and G have tied ranks at 4th and 5th: the average is 4.5.

For Perry, B, G, H and I have tied ranks at 4th, 5th, 6th and 7th: the average is 5.5.

K rank	1	9	4.5	10	8	3	4.5	2	6	7
P rank	1	5.5	3	8	9	2	5.5	5.5	5.5	10
d	0	3.5	1.5	2	1	1	1	3.5	0.5	3
d^2	0	12.25	2.25	4	1	1	1	12.25	0.25	9

> **Note:** tied ranks do not always result in 0.5s.

$\Sigma d^2 = 43$

There are 10 drinks so $n = 10$.

$$r_s = 1 - \frac{6 \times 43}{10(100 - 1)}$$

$$= 1 - \frac{258}{990}$$

$$= 1 - 0.261 = 0.739$$

A rank correlation coefficient of 0.739 suggests that there is fairly strong agreement between Kevin and Perry's taste in drinks.

Exercise 6H

1. Which of the numbers $^-0.73$, 0.29 and 0.87 indicates the least correlation? Give a reason for your answer.

2. From the numbers 0, 0.78, $^-0.39$, 0.12, $^-0.81$ and 1.32 choose the most likely correlation coefficient that matches each of the scatter graphs A, B and C.

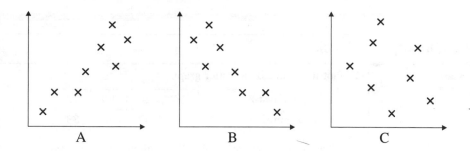

A   B   C

Which one of the numbers cannot be a rank correlation coefficient? Give a reason for your answer.

3. The table below gives the marks out of 20 obtained by five students in two tests.

Test 1	2	7	8	9	13
Test 2	4	5	7	10	14

Work out the value of Spearman's rank correlation coefficient between tests 1 and 2. What does this tell you about the students and the tests?

4. Shoshana and Simon ranked seven sports in order of preference.

Sport	Tennis	Squash	Hockey	Football	Badminton	Rugby	Swimming
Shoshana	1	5	4	6	3	7	2
Simon	3	6	4	2	7	5	1

Calculate Spearman's rank correlation coefficient. Comment on your answer.

5. Earl kept a record of how far he travelled, in km, and the petrol used, in litres.

Distance (km)	480	592	656	672	704	720
Petrol (litres)	35	46	44	45	52	52

Calculate Spearman's rank correlation coefficient. Comment on your answer.

6. The following table gives the times taken, in minutes, and the number of errors made in a typing test.

Typist	A	B	C	D	E	F	G	H	I
Time	15	16	10	19	23	15	20	21	15
Errors	7	8	5	6	3	5	4	2	9

Calculate Spearman's rank correlation coefficient. Comment on your answer. Which typist would you employ? Give a reason for your answer.

Summary

You should now be able to	Check out 6
1 Draw and interpret a scatter diagram.	1 Plot these data on a scatter graph. 	% in test X	40	64	36	50	72	 	% in test Y	52	75	47	62	83	 What does the scatter graph tell you about the two tests?
2 Draw a line of best fit.	2 Draw a line of best fit on your scatter graph.
3 Find the equation of a line of best fit.	3 Calculate the equation of your line of best fit.
4 Calculate and interpret Spearman's coefficient of rank correlation.	4 Six different jams were ranked by two judges. 	Judge 1 rank	1	2	3	4	5	6	 	Judge 2 rank	5	3	2	6	4	1	 Calculate Spearman's coefficient of rank correlation. What does your answer tell you about the opinions of the two judges?

Revision Exercise 6

1. (a) There is a high degree of negative correlation between the amount of heat escaping through the roof of a house and the thickness of the insulating material used in the loft.
  Explain what is meant by this statement.

  (b) There is a high degree of positive correlation between the temperature in New Zealand and the amount of coal sold in the United Kingdom.
  Explain what is meant by this statement.

  (c) In only one of the above cases is there a direct causal relationship between the two variables. In which case is there a direct causal relationship?

  (d) Give an example of your own where there is a high degree of positive correlation between two variables and a direct causal relationship.   [NEAB]

2. The table on the next page gives the age, in years, and the 'nearest vision distance', in centimetres, of each of ten people. The nearest vision distance is the closest distance at which a person can read.

Person	A	B	C	D	E	F	G	H	I	J	Mean value
Age (years)	20	25	30	35	40	45	50	55	60	65	
Nearest vision distance (cm)	10	21	18	24	28	30	34	40	45	50	30

(a) Draw a scatter diagram of this information.
(b) Calculate the mean age of the 10 people.
(c) (i) Plot the point on the scatter diagram to show the mean age and mean 'nearest vision distance'. Label this point M.
   (ii) Draw a straight line to fit your points.
(d) Use your line to estimate the 'nearest vision distance' of a person
   (i) aged 15,
   (ii) aged 57.
(e) Which of your two estimates do you think is the more reliable? Explain your reasoning.
(f) Which person appears to have an unusually high 'nearest vision distance'? [NEAB]

3. The time (in seconds) taken by eight boys to solve 10 addition sums and 10 multiplication sums is given in the following table.

Boy / Operation	A	B	C	D	E	F	G	H
Addition time (s)	48	20	13	18	40	29	34	43
Multiplication time (s)	53	26	17	20	38	40	37	49

(a) Plot a scatter graph of these data.
(b) (i) Calculate the mean of the addition times.
   (ii) Calculate the mean of the multiplication times.
(c) Draw the line of best fit on your graph.
(d) Use your line to estimate the time it would take for a boy to complete the ten multiplication sums if he took
   (i) 25 seconds to complete the ten addition sums,
   (ii) 60 seconds to complete the ten addition sums.
(e) Which of these answers is the most reliable? Give a reason. [SEG]

4. Brunel plc is keen to set up a forecasting system which will enable them to estimate maintenance costs for delivery vehicles of various ages.
The following table summarizes the age in months (x) and maintenance costs £ (y) for a sample of ten such vehicles.

Vehicle	A	B	C	D	E	F	G	H	I	J
Age, months (x)	63	13	34	80	51	14	45	74	24	82
Maintenance cost, £ (y)	141	14	43	170	95	21	72	152	31	171

(a) Draw a scatter diagram of this data.

(b) Find the mean value of the ages (x) and maintenance cost (y)

(c) Use your results from (b) and the fact that the line of best fit for the data passes through the point (20, 24.5) to draw this line on the graph.

(d) Estimate from your line the maintenance cost for a vehicle aged
    (i) 85 months,   (ii) 5 months,   (iii) 60 months.

(e) Order these forecasts in terms of their reliability. Justify your choice.   [NEAB]

5. The scatter diagram shows the weights, in kilograms, and the heights, in centimetres, of seventeen adult males in a rugby club.

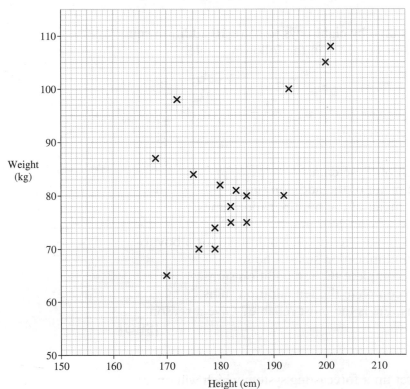

Heights and weights of males in a rugby club

(a) What is the weight of the heaviest player?

(b) What is the height of the shortest player?

(c) One of the players is particularly heavy for his height. Write down the coordinates of the **X** that represents this player.

(d) What type of correlation does the diagram show?

(e) The team captain uses the diagram below to categorize the players.

		Height	
		Very tall	Not very tall
Weight	Very heavy	3	2
	Not very heavy	1	11

How does the team captain decide if someone is
(i)   very heavy,   (ii)   very tall?   [NEAB]

6. The dots on the scatter diagram represent the sale price and the usual price of some cameras sold at the Nixon Camera Company.

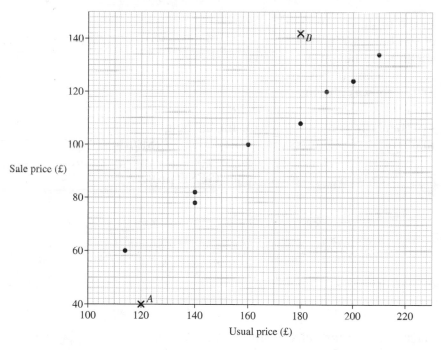

The average price of these cameras before the sale was £167.
The average price of these cameras in the sale was £101.

(a) Copy the scatter diagram and plot the point representing these average prices.

(b) On your scatter diagram, draw the line of best fit.

(c) Use your line to estimate the saving you would make on a camera whose usual price was £150.

The Beetall Camera Company recorded similar information in its sale. The line of best fit for this company was a line joining the points *A* and *B* on the scatter diagram.

Dawn has £55 to spend and her sister Julie has £120 to spend. They both want to buy a camera.

(d) (i)   Which shop would give Dawn a better deal?
Give a reason for your answer.

(ii)   Explain why Julie would get a better deal at the other shop. [SEG]

7. The jump height of a boy is the height above the floor that he can jump up to and touch.
The table gives the heights of eight boys and their jump heights.

	A	B	C	D	E	F	G	H
Height of boy (cm)	152	165	144	169	154	151	160	157
Jump height (cm)	210	235	208	252	222	224	230	241

(a) Plot a scatter graph of these data.
The mean height of the boys is 156.5 cm.
(b) Calculate the mean jump height of the boys.
(c) On your graph, draw a line of best fit.
(d) A boy of height 163 cm was absent.
Use your line of best fit to predict his jump height.
(e) One boy has a particularly good jump height for his height.
Which boy is that? [SEG]

8. Tanya produced these three scatter diagrams in her GCSE projects.

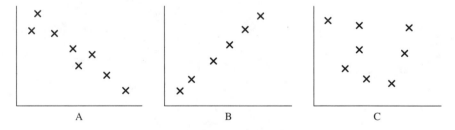

The titles for the diagrams are reproduced on the table below.
Copy and complete this table to show which diagram would best suit the titles and indicate the type of correlation shown by each diagram. [SEG]

Title	Diagram letter	Type
Marks scored in a test and the percentage for that test.		
The age of a family car and its value.		
The time to run 200 metres and the number of people in their family.		

9. The graph shows the line of best fit to show the relationship between the straight-line distance (x) and the journey distance (y) of pupils to school.

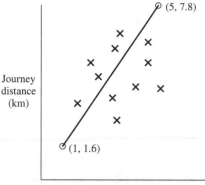

(a) Calculate the equation of the line of best fit.
(b) Explain why the intercept with the y-axis is small.
(c) Interpret the gradient of the line.   [SEG]

10. The following data relate to the age and weight of ten randomly chosen children in Bedway Primary School.

Age (years)	7.8	8.1	6.4	5.2	7.0	9.9	8.4	6.0	7.2	10.0
Weight (kg)	29	28	26	20	24	35	30	22	25	36

(a) Draw a scatter diagram to show this information.

The mean age of this group of children is 7.6 years.

(b) Calculate the mean weight of this group.
(c) On the graph, draw the line of best fit.
(d) Use your graph to find the equation of this line of best fit, in the form $y = mx + c$.

Jane is a pupil at Bedway Primary School and her age is 8.0 years.

(e) Use your answer to (d) to estimate Jane's weight.
(f) Give **one** reason why a prediction of the weight of a twelve-year-old from your graph might not be reliable.   [SEG]

11. A student carried out an experiment to investigate the effectiveness of a fertilizer.
Five plants of the same height were chosen and a different amount of fertilizer was given to each plant. The student plotted the results of the experiment on a scatter diagram and drew the line of best fit.
(a) Comment on the line of best fit.
(b) Explain why it would be inadvisable to use this line of best fit to predict the growth for a 6-gram dose of fertilizer.

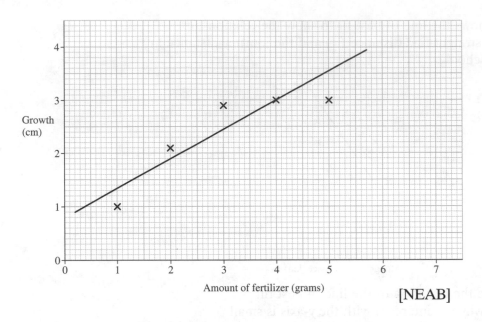

Amount of fertilizer (grams)

[NEAB]

12. For twelve consecutive months a factory manager recorded the number of items produced by the factory and the total cost of their production.

Number of items (x)	18	36	45	22	69	72	13	33	60	79	10	53
Production cost (£ y)	37	54	63	42	84	91	33	49	79	98	32	70

(a) Draw a scatter diagram for the data.
(b) Find the mean value of x and y.
(c) Use your results from (b), and the fact that the line of best fit for the data passes through the point (70, 88), to draw this line on the graph.
(d) Estimate, from your line, the production cost for
   (i)   65 items,
   (ii) 84 items.
(e) Which of your forecasts in (d) is the more reliable? Justify your choice.
(f) Find, from your graph, the gradient of the line.
(g) Describe briefly, in the context of the question, what the gradient measures.   [NEAB]

13. The table shows the engine capacity (x) and the selling price (y) of 6 different models of car.

Capacity of engine (cc)	1500	1300	1100	1600	1250	1800
Selling price (£)	9800	8200	7000	11 200	7800	12 000

(a) Draw a scatter diagram to show this data.

The mean selling price of these cars is £9370.

(b) Construct a line of best fit through the points on the diagram.
(c) Obtain the equation of this line in the form $y = mx + c$.
(d) (i)   Use this information to predict the selling price of a car
    with an engine capacity of 1700 cc.
   (ii)   Give **one** reason why this answer may not be reliable.

[SEG]

14. When two judges ranked ten television programmes, the value of
Spearman's rank correlation coefficient was −0.05.
(a) Which of the following scatter diagrams could represent this
    information?

(b) Give estimates for Spearman's rank correlation coefficient for
    the other two diagrams.   [SEG]

15. The table shows the number of points achieved and the number of
goals conceded by eight netball teams.

Team	A	B	C	D	E	F	G	H
Points	29	28	27	25	24	22	23	21
Goals conceded	15	12	13	16	19	14	21	17

(a) Calculate Spearman's coefficient of rank correlation between
    the number of points achieved and the number of goals
    conceded.

$$\text{Spearman's coefficient} = 1 - \frac{6 \sum d^2}{n(n^2 - 1)}$$

Spearman's coefficient of rank correlation between the number of
points achieved and the number of goals **scored** is 0.45.

(b) Which is the better relationship to use for predicting the points
    obtained, the relationship between:
    (i)   the number of goals conceded and the number of points
           achieved, or
    (ii)   the number of goals scored and the number of points
           achieved?
    Explain why you chose this answer.   [SEG]

16. Two judges had to rank ten gymnasts in order of ability. The table gives their respective rankings.

Gymnast	A	B	C	D	E	F	G	H	I	J
Judge 1	5	1	10	6	2	7	9	8	3	4
Judge 2	6	4	10	9	1	3	8	5	7	2

(a) Use the table to calculate Spearman's rank correlation coefficient between the judges' rankings.
(b) What does your answer suggest about the judges' level of agreement? [SEG]

17. At the Deepdale 'Best of British Pie' competition two judges award marks for nine different pies as follows.

Pie	A	B	C	D	E	F	G	H	I
Judge 1	18	24	23	13	27	19	30	10	20
Judge 2	7	18	9	4	17	8	20	5	10

(a) What do the scores tell you about the two judges?
(b) (i) Calculate Spearman's coefficient of rank correlation between the two judges.
(ii) What does your result suggest about the judges' decisions? [SEG]

18. The table shows the results of three tests given to seven pupils in a technology class.

Pupil	Test 1 Class position	Test 2 Score out of 60				Test 3 percentage
A	1st	40				*
B	3rd	35				48
C	2nd	34				40
D	7th	25				36
E	5th	29				38
F	6th	20				*
G	4th	42				53

(a) Copy the table and use it to calculate Spearman's coefficient of rank correlation between Test 1 and Test 2.

All seven pupils took Test 3.
The rank order for Test 3 was the same as the rank order for Test 2.
The percentages of pupils **A** and **F** in Test 3 were then lost.

(b) What would be the maximum percentage in Test 3 for pupil **F**?

(c) What would be
   (i)   the lowest percentage in Test 3 for pupil **A**?
   (ii) the highest percentage in Test 3 for pupil **A**?

(d) Without further calculation, state the value of the rank
    correlation coefficient between Test 2 and Test 3.   [SEG]

19. In a music festival, each competitor is judged on his performance
on two different musical instruments. The judge awards marks out
of 100 for each instrument, as follows.

Competitor	A	B	C	D	E	F
1st Instrument	90	75	62	70	75	56
2nd Instrument	95	76	64	76	86	60
Rank 1	1					
Rank 2	1					

(a) Copy the table and complete the ranks.

The rank correlation coefficient for these data was found to be 0.96.

It was later discovered that the marks from one of the judges, for
one competitor, had been misread. This competitor should have
had 10 more marks on his second instrument.

The mark was changed and on recalculation it was found that the
correlation coefficient remained the same at 0.96.

(b) (i)   Which competitor's mark was originally incorrect?
    (ii) Give a reason for your answer.   [SEG]

20. The table shows the percentage of the total lottery ticket sales and
the National Lottery grant for each of the ten different regions.

Region	Percentage of total lottery sales	Lottery grants (£m)	Ranks Sales	Ranks Grants	d	d^2
London	22.3	125.0				
Midlands	15.7	11.7				
North West	11.7	5.7				
The South	10.4	17.5				
Yorkshire	9.5	5.1				
Scotland	8.8	5.9				
Wales	7.4	4.3				
The East	6.5	12.6				
North East	5.3	2.4				
N. Ireland	2.4	0.5				

(a) Copy the table and complete the ranking columns for each set of data.

The formula for calculating Spearman's rank correlation coefficient is:

$$r_s = 1 - \frac{6\sum d^2}{n(n^2 - 1)}$$

(b) Calculate this rank coefficient for the above data.
(c) Use your value of Spearman's rank correlation coefficient to comment about the way money is distributed.
(d) By what amount could the National Lottery grant for London be reduced without changing the value of Spearman's rank correlation coefficient?   [SEG]

21. Attendances at some of the most popular tourist attractions in the UK charging for admission (in millions) are shown below.

Attractions	1981	1991
Madame Tussaud's	2.0	2.2
Alton Towers	1.6	2.0
Tower of London	2.1	1.9
Natural History Museum	3.7	1.6
Chessington World of Adventures	0.5	1.4
Science Museum	3.8	1.4
London Zoo	1.1	1.1

(a) Calculate to 3 decimal places, Spearman's rank correlation coefficient for these data.
(b) The equivalent rank correlation coefficient for 1986 attendances compared with 1991 was +0.724.

By reference to this value and the one obtained in part (a), comment on the differences apparent.   [NEAB]

22. In a ski jumping contest each competitor made two jumps. The orders of merit for the fifteen competitors who completed both jumps are shown in the following table:

Ski jumper	A	B	C	D	E	F	G	H	I	J	K	L	M	N	O
First jump	8	10	15	3	9	2	11	1	12	4	5	13	6 =	6 =	14
Second jump	5	15	6	1	12	2	14	3	7 =	10	11	7 =	4	13	7 =

(a) Calculate Spearman's rank correlation coefficient for these data.

(b) What does this tell us about the skiers' performances on the two jumps?

(c) State what you would conclude if from three further sets of rankings, involving the same number of skiers, the following values for Spearman's rank correlation coefficient were obtained.

   (i) −0.04   (ii) 0.92   (iii) −1.14   [NEAB]

23. The table below shows the weather recorded on Monday 14 March 1997 in ten Scottish towns.

WEATHER

Last night's reports for 24 hours to 6 pm

	Sunshine hours	Maximum temperature °F	Weather (day)	Rank sunshine hours	Rank temperature		
Aberdeen	2.7	36	snow				
Aviemore	1.8	32	cloudy				
Edinburgh	0.2	33	cloudy				
Eskdalemuir	0.0	34	cloudy				
Glasgow	0.2	39	cloudy				
Kinloss	2.6	37	bright				
Lerwick	3.4	35	snow				
Leuchars	5.7	38	bright				
Tiree	1.2	40	bright pm				
Wick	6.4	37	snow am				

Source: The Guardian, 15 March 1997

(a) Rank each town in relation to hours of sunshine and maximum temperature °F.

(b) Using the formula $1 - \dfrac{6 \sum d^2}{n(n^2 - 1)}$ calculate Spearman's rank correlation coefficient for these data.

(c) Comment on the correlation shown.

(d) Suppose the equivalent rank correlation coefficients for Saturday 12 March and Sunday 13 March were, respectively, +0.07 and −0.95.

What would these two values suggest about the relationship between daily hours of sunshine and maximum temperature?

   (i) Saturday value   (ii) Sunday value   [NEAB]

7 Probability

Probability is a measure of chance:

What are the chances of that happening, eh?

This unit will show you how to

✦ Work out simple probabilities
✦ Draw sample space and tree diagrams
✦ Calculate probabilities for exclusive and independent events
✦ Carry out simulation
✦ Use the discrete uniform and binomial distributions

Before you start

You need to know how to	Check in 7
1 Convert from fractions to decimals to percentages	**1** Copy and complete the table.

Fraction	Decimal	Percentage (%)
$\frac{2}{5}$		
	0.35	
		18

2 Add, subtract, multiply and divide fractions and decimals

2 (a) Calculate: (b) Work out:
(i) $\frac{3}{4} + \frac{5}{6}$ (i) 0.3×0.4
(ii) $\frac{3}{4} \times \frac{8}{9}$ (ii) $0.5 \div 0.08$
(iii) $\frac{5}{6} \div \frac{2}{3}$

3 Draw a line graph

3 On graph paper, draw straight lines to connect the following coordinates in alphabetical order:
A(2, 2),   B(4, 4),   C(6, 2), D(7, 6),   E(8, 5)

7.1　Theoretical probability

To cope in an increasingly uncertain world, it helps if you understand the laws of chance.

Probability is a measure of likelihood.

In probability, an **event** is a situation, such as rolling a dice, or winning the FA Cup in football. An **outcome** is a particular result of an event, such as scoring a 4 on a dice, or Arsenal winning the FA Cup.

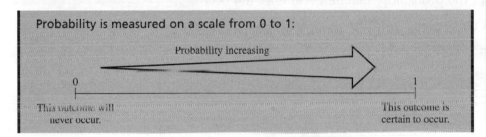

Probability is measured on a scale from 0 to 1:

Probability increasing

0 — This outcome will never occur.

1 — This outcome is certain to occur.

Probabilities are expressed using either fractions, decimals or percentages.

If there is a 10% chance of rain tomorrow, this could be written:

$$P(R) = 0.1$$

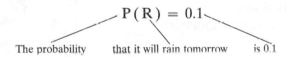

The probability　that it will rain tomorrow　is 0.1

● Example

Match each of the statements to position A, B or C on the probability line.

0　　　　　　　　　　　　　　　　　　　　　　1
A　　　　　　　　　B　　　　　　　　　C

1.　There will be a school trip to the moon next week.
2.　You will drink something today.
3.　It will be sunny next Saturday.

In each case give a reason.

Statement 1 matches to A because it is highly unlikely.
Statement 2 matches to C because it is highly likely.
Statement 3 matches to B because it is possible.

Probability of an event

An event may have a number of possible outcomes. If they are all
equally likely to occur, then you can use this formula to work out
probability:

> Probability of an event $= \dfrac{\text{number of outcomes leading to an event}}{\text{total number of all possible outcomes}}$

For example, the probability of drawing a club from a pack of playing
cards is

$$\dfrac{13}{52}$$

— number of clubs
— number of cards

> Any card has an equal
> chance of being drawn and
> so the outcomes are equally
> likely.

You can also work out the probability that an event **does not** occur.

For example, if there is a $\frac{1}{10}$ chance of rain tomorrow, then there is a $\frac{9}{10}$
chance that it will not rain tomorrow.

> The probability that an event A does not occur is usually written P(A$'$).

> P(A$'$) = 1 − P(A) for any event A.

● **Example**

A fair six-sided dice is thrown.

(a) Find the probability of throwing a multiple of 3.
(b) Find the probability of **not** throwing a square number.

(a) There are six possible outcomes: 1, 2, 3, 4, 5 and 6.
    There are two multiples of 3:   3 and 6.
    P (a multiple of 3) $= \frac{2}{6} = \frac{1}{3}$

(b) There are two square numbers: 1 and 4.
    P (a square number) $= \frac{2}{6}$
    P (not a square number) $= 1 - \frac{2}{6}$
                         $= \frac{4}{6} = \frac{2}{3}$

Exercise 7A

1. A bag contains 4 red, 3 blue, 1 green and 2 yellow marbles.
    A marble is removed at random.
    What is the probability that it is

(a) red,   (b) yellow or blue,
(c) not red,   (d) not yellow or red?

2. A complete pack of playing cards is shuffled.
A card is selected at random.
What is the probability that it is

(a) a king,
(b) a red queen,
(c) not a red queen,
(d) a picture card,
(e) not a queen or a king?

3. At a bottle stall, tickets number from 1 to 500 are put in a bag.
Tickets ending in 25, 50, 75, 00 win the big prize.
A ticket is selected at random.
Find the probability that the selected ticket wins the big prize.

4. A regular pentagonal spinner has the numbers 1, 2, 3, 4, 5 marked
on its edges. What is the probability of it landing on

(a) an even number,
(b) a prime number,
(c) not a multiple of 3?

5. A CD has 12 tracks.

4 recordings are by female groups.
3 recordings are by male groups.
3 recordings are by female solo artists.
2 recordings are by male solo artists.

The CD player is set to random.
What is the probability that

(a) a song by a solo artist is selected,
(b) a song not by a solo female artist is selected?

6. Five children, Alan, Ann, Kate, Simon and Thomas, are playing
musical chairs.

(a) Explain why the probability of Alan losing the first round is not 0.2.

Five pieces of string, all of different lengths, are selected at random
by the five children.

(b) What is the probability that a boy selects the shortest piece of string?
(c) What is the probability that a child whose name begins with the
letter A selects the longest piece of string?

7.2 Experimental probability

When you throw a dice you usually assume that it is fair.

However, this is an **assumption** and it is not necessarily true.

> You can estimate the probability of an event by physically trying it out.

Example

Gaius buys seeds of a particular flower in packets of 100. Sometimes they germinate and sometimes they do not. He wants to find the **experimental probability** that a seed will germinate. He records the results for five packets in a table:

Packet	1	2	3	4	5
Number of seeds that germinated	70	30	60	50	60

Find the proportion, or **relative frequency**, of seeds that germinated. Draw a frequency polygon to illustrate the relative frequency. Use it to estimate the probability of germination.

Redraw the table using the cumulative totals:

Hint:
You can find out more about frequency polygons on page 70.

Number of seeds sown	Number of seeds germinated	Relative frequency
100	70	0.7
200	100	0.5
300	160	0.53
400	210	0.525
500	265	0.53

Relative frequency

$= \dfrac{\text{number of seeds germinated}}{\text{number of seeds sown}}$

Draw a graph of the results.

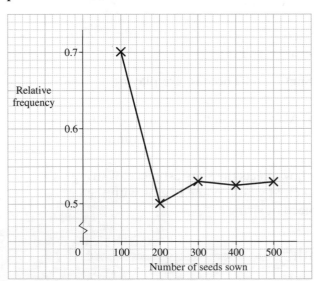

The graph shows that the relative frequency is changing less as the number of seeds increases. It is settling down to around 0.53. This is a good estimate of the probability of germination.

This is a **statistical experiment**, and it involves the repetition of a number of **trials**.

> **Relative frequency** is the proportion of 'successful' trials in an experiment.

$$\text{Relative frequency} = \frac{\text{number of successful trials}}{\text{total number of trials}}$$

Note: The larger the number of trials, the closer the relative frequency is likely to be to the true probability.

Exercise 7B

1. Graham is conducting a statistical experiment on a biased coin. He has recorded the amount of heads for differing numbers of throws:

Number of throws	50	100	150	200	250	300
Number of heads	26	48	65	76	102	118
Relative frequency	0.52					

   (a) Complete the table for relative frequency.
   (b) On graph paper, draw a frequency polygon of relative frequency against number of throws.
   (c) Give an estimate for the probability of obtaining a head.

2. It is known that a biased dice has a probability of 0.2 of getting a six.

   (a) How many sixes would you expect to get in 100 throws?
   (b) How many times would you expect **not** to get a six in 500 throws?

3.

When I play a game using a fair dice, why is it that I seem to get more 3s than 6s?

Perform an experiment to see if this is the case.

4. It is known that 1.5% of patients react adversely when treated by a particular drug.
800 patients are treated.
How many of these would you expect to react adversely?

7.3 Sample space diagrams

When you want to find the probability of two or more events occurring you can list the outcomes on a **sample space diagram**.

Example

Two fair dice are rolled.

The score is the difference between the numbers shown on the dice.

(a) Draw a sample space diagram showing all the possible outcomes.
(b) Find the probability of getting a score of 3.

(a) The possible outcomes are listed in the table:

This is the difference between 1 and 2

—	1	2	3	4	5	6
1	0	1	2	③	4	5
2	1	0	1	2	③	4
3	2	1	0	1	2	③
4	③	2	1	0	1	2
5	4	③	2	1	0	1
6	5	4	③	2	1	0

Note the patterns in the table – it makes the data easier to follow.

(b) There are 36 equally likely outcomes.
Six of these result in a score of 3.
The probability of a score of 3 is $\frac{6}{36} = \frac{1}{6}$.

You can draw sample space diagrams in simple cases of three events as well.

Example

Three fair coins are tossed and the possible results of each coin are either 'heads' (H) or 'tails' (T).

(a) List all of the possible outcomes.
(b) Find the probability that two heads are tossed.

(a) The sample space diagram shows all eight possible outcomes:

HHH	HHT	HTH	THH
TTT	TTH	THT	HTT

(b) There are 8 possible outcomes.
There are 3 outcomes containing two heads.

P (two heads) $= \frac{3}{8}$

Exercise 7C

1. Two fair dice are rolled. The score is the sum of the scores on each dice. Draw a sample space diagram to show all the possible outcomes. Use the diagram to find the probability that the total score is

   (a) 4,   (b) 11
   (c) What total is most likely?
   (d) What is the probability of this score?
   (e) What is the probability that the two dice show the same score?

2. A regular pentagonal spinner has one of the numbers 1, 2, 3, 4, 5 on each edge. The spinner is spun twice and the score is the product of the two scores. Draw a sample space diagram to show all the possible outcomes. Use the diagram to find the probability that the total score is

   (a) 6,   (b) 11
   (c) What score is most likely?
   (d) What is the probability of this score?

3. A regular tetrahedron has the numbers 1, 2, 3 and 4 on each face. A second regular tetrahedron has the numbers 2, 4, 6 and 8 on each face. The score is the sum of the scores on the two tetrahedra. Draw a sample space diagram to show all the possible outcomes. Use the diagram to find the probability that the total score is

   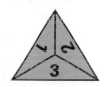

   (a) 4,   (b) 11
   (c) Why is it necessary for the tetrahedra to be regular?

4. Two dice are rolled and the product of the two scores is recorded

   (a) Draw a sample space diagram to illustrate the scores.
   (b) Use your diagram to obtain the probability that the score was
        (i) 10,   (ii) an odd number.

5. Two fair pentagonal spinners, one with the numbers 1, 3, 5, 7, 9 and the other with the numbers 2, 4, 6, 8, 10 are spun and the difference in the scores is recorded.

   (a) Draw a sample space diagram to illustrate the scores.
   (b) Use your diagram to obtain the probability that the score was
        (i) 7,   (ii) 5   (iii) What is the most likely score?

6. A regular tetrahedron has the numbers 1, 2, 3, 4 on its faces. A fair dice has the numbers 1, 2, 3, 4, 5 and 6 on its faces. The two scores are added together.

   (a) Draw a sample space diagram to illustrate the possible total scores. What is the probability that the total score is
   (b) 4,   (c) more than 7?
   (d) Why does the tetrahedron have to be regular?

7.4 Mutually exclusive events

Two events are **mutually exclusive** if they cannot occur at the same time.

For mutually exclusive events A and B:

$P(A \text{ or } B) = P(A) + P(B)$

This is known as the addition law for mutually exclusive events.

Note: This is sometimes called the OR rule.

Note: the rule can be extended for 3 or more mutually exclusive events – you just add up the probabilities.

Example

A particular football team, Cottford Athletic, reckon they will win (W), draw (D) or lose (L) their next league game with these probabilities:

$$P(W) = 0.5, P(D) = 0.3, P(L) = 0.2$$

Find the probability that they will not lose.

Only one of the outcomes can occur in any one match so they are mutually exclusive.

$$
\begin{aligned}
P(L') &= P(W \text{ or } D) \\
&= 0.5 + 0.3 \\
&= 0.8
\end{aligned}
$$

Note that if you add up all three probabilities you get 1. This means that Cottford Athletic either win, draw or lose; there are no other possibilities.

Remember:
L' means : not L.

The sum of the probabilities of all mutually exclusive events is 1.

A set of events is **exhaustive** if at least one of them must occur.

Example

A fair dice is thrown.

A is the event 'obtaining an even number'.
B is the event 'obtaining a multiple of 3'.
Decide whether events A and B are
(a) mutually exclusive,   (b) exhaustive.

(a) A includes the outcomes {2, 4, 6}.
   B includes the outcomes {3, 6}.

   A and B are not mutually exclusive because they both contain the outcome 6.

(b) A and B are not exhaustive because they do not include the outcomes 1 or 5.

Exercise 7D

1. A box contains coloured marbles. The number of each colour is given in the table.

Colour	Red	Blue	Green	Yellow
Number of marbles	35	30	20	15

A marble is selected at random.
What is the probability that the marble is

(a) red,

(b) red or blue,

(c) red and blue,

(d) not red?

2. In the game of Bingo, 90 discs are numbered consecutively from 1 to 90. A disc is selected at random. What is the probability that the disc

(i)   is the number 37,

(ii)   ends with a zero,

(iii) is a multiple of 7,

(iv)   is a multiple of 8,

(v)   is a multiple of 7 or 8,

(vi)   is a multiple of 7 and 8,

(vii) is a multiple of 3,

(viii) is not a multiple of 3?

3. In a small school, a class consists of children of a variety of ages as given in the table.

5-year-old girls	5-year-old boys	6 year-old girls	6-year-old boys	7-year-old girls	7-year old boys
3	4	6	8	5	2

A pupil is selected at random.
What is the probability that the pupil is

(i)   a girl,

(ii)   not 5 years old,

(iii) a boy and 6 years old,

(iv)   a girl or 6 years old,

(v)   6 or 7 years old,

(vi)   6 and 7 years old?

7.5 Independent events

Two events are **independent** if the outcome of one does not affect the outcome of the other.

> If A and B are two independent events, then:
>
> $P(A \text{ and } B) = P(A) \times P(B)$
>
> This is known as the **multiplication** law for independent events.

Note: This law is sometimes called the AND rule.

You can extend the rule for 3 or more independent events – just multiply the probabilities together.

Example 1

An aircraft is powered by two engines, A and B, which operate independently of each other.

The probability that A malfunctions is 0.01.

The probability that B malfunctions is 0.02.

The aircraft will fly as long as at least one of the engines is working.

(a) What is the probability that both engines are malfunctioning at the same time?

(b) What is the probability that the aircraft will be able to fly?

(a) $P(\text{A and B malfunctioning}) = 0.01 \times 0.02$
$$= 0.0002$$

(b) $P(\text{aircraft able to fly}) = 1 - 0.0002$
$$= 0.9998$$

Example 2

Katrina is aiming three darts at a dartboard. The probability that she will hit the central region (the bulls-eye) with a single dart is 0.2.

The outcome of a particular throw does not affect her performance in the next throw.

What is the probability that she hits the bulls-eye with exactly one dart out of the three?

Let H be the event 'Katrina hits the bulls-eye with a dart'.

Then H′ is the event 'Katrina misses the bulls-eye with a dart', and so on.

$$P(H) = 0.2 \quad \text{and} \quad P(H') = 1 - 0.2 = 0.8$$

$P(\text{Katrina only succeeds with the first dart}) = P(\text{H and H}' \text{ and H}')$
$$= 0.2 \times 0.8 \times 0.8$$
$$= 0.128$$

$P(\text{Katrina only succeeds with the second dart}) = P(\text{H}' \text{ and H and H}')$
$$= 0.8 \times 0.2 \times 0.8$$
$$= 0.128$$

$P(\text{Katrina only succeeds with the third dart}) = P(\text{H}' \text{ and H}' \text{ and H})$
$$= 0.8 \times 0.8 \times 0.2$$
$$= 0.128$$

$P(\text{Katrina only succeeds with exactly one dart})$
$$= P(\text{first dart or second dart or third dart})$$
$$= 0.128 + 0.128 + 0.128$$
$$= 0.384$$

Hint: When you are working out a difficult problem in probability, it helps to use letters to represent the events.

This example could be done using a tree **diagram**. These are explained in the next section.

Exercise 7E _____

1. A fair coin is tossed and a dice with the numbers 2, 4, 6, 8, 10 and 12 is rolled.
   Draw a sample space diagram to show all the outcomes.
   (a) What is the probability that a head is tossed **and** a 4 is rolled?
   (b) What is the probability that a head is tossed **or** a 4 is rolled?

2. An archer shoots at a target. The probability of hitting the gold area is 0.2. He fires two shots at the target.
   (a) What is the probability that both arrows hit the gold area?
   (b) What is the probability that exactly one arrow hits the gold area?

   Another archer has a probability of 0.4 of hitting the gold area.
   He fires three shots at the target.
   (c) What is the probability that all three shots miss the gold area?
   (d) What is the probability that at least one shot hits the gold area?

3. Three children take a test. The probability that Chris passes is 0.8, the probability that Georgie passes is 0.9 and the probability that Phil passes is 0.7.
   (a) What is the probability that all three pass?
   (b) What is the probability that all three fail?
   (c) What is the probability that at least one passes?

4. The probability that Sharon beats George in a game of tennis is 0.6. They play three games.
   What is the probability that
   (a) Sharon wins all three games,
   (b) George wins all three games,
   (c) they both win at least one game?

5. John drives to work and passes three sets of traffic lights.
   The probability that he has to stop at the first is 0.6.
   The probability that he has to stop at the second is 0.7.
   The probability that he has to stop at the third is 0.8.
   (a) Calculate the probability that he stops at all three sets of traffic lights.

   He arrives late if he has to stop at any two sets of traffic lights.
   (b) Calculate the probability that he is late.

7.6   Tree diagrams

Tree diagrams can help you to work out the probability of combined events.

Tree diagrams with replacement

Example

A bag contains 10 red marbles and 5 blue marbles. A marble is drawn at random, its colour is noted and it is then replaced in the bag. A second marble is then drawn out and its colour is noted.

(a) Illustrate this situation with a tree diagram.

(b) Find the probability that both marbles are the same colour.

(a)

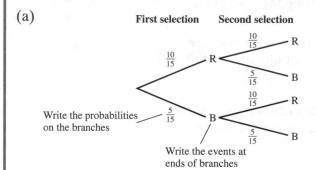

R = red marble
B = blue marble

Write the probabilities on the branches

Write the events at ends of branches

> Tree diagrams display all possible outcomes in a situation.

(b) To find the probability that both marbles are of the same colour, follow the 'paths' indicated in blue on the tree diagram:

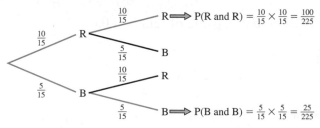

$$P(R \text{ and } R) = \tfrac{10}{15} \times \tfrac{10}{15} = \tfrac{100}{225}$$

$$P(B \text{ and } B) = \tfrac{5}{15} \times \tfrac{5}{15} = \tfrac{25}{225}$$

> You **multiply** the probabilities along a given path.

P (both red **or** both blue)

$= \tfrac{100}{225} + \tfrac{25}{225}$

$= \tfrac{125}{225}$

$= \tfrac{5}{9}$

> You **add** the probabilities for mutually exclusive paths.

Tree diagrams without replacement

Sometimes the events illustrated by a tree diagram are not independent. The outcome of the first event may affect subsequent events.

Example

A box contains 12 beads. Five are yellow and the rest are green. A bead is removed from the box and its colour is noted. It is not returned to the box. A second selection is then made and the process is repeated, followed by a third selection.

(a) Draw a tree diagram outlining this situation.

(b) Find the probability of selecting exactly two green beads.

(a)

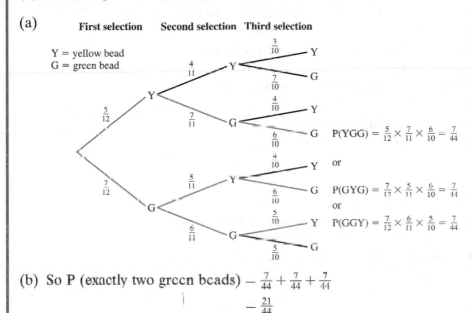

First selection   Second selection   Third selection

Y = yellow bead
G = green bead

$P(YGG) = \frac{5}{12} \times \frac{7}{11} \times \frac{6}{10} = \frac{7}{44}$

or

$P(GYG) = \frac{7}{12} \times \frac{5}{11} \times \frac{6}{10} = \frac{7}{44}$

or

$P(GGY) = \frac{7}{12} \times \frac{6}{11} \times \frac{5}{10} = \frac{7}{44}$

(b) So P (exactly two green beads) $= \frac{7}{44} + \frac{7}{44} + \frac{7}{44}$

$= \frac{21}{44}$

Hint: At the second selection there are only 11 beads left. At the third selection there are only 10.

Note: This tree diagram shows three stages of branches. Too many stages can be impractical with this type of diagram.

Remember:
And is ×   Or is +

Exercise 7F

1. A bag contains 3 red and 2 blue discs.
    A disc is selected at random, its colour noted, and replaced in the bag.
    A second disc is selected at random.
    Draw a tree diagram to show all the possible outcomes.

    (a) What is the probability that both discs are red?
    (b) What is the probability that both discs are blue?
    (c) What is the probability that the discs are of the same colour?
    (d) What is the probability that the discs are of different colours?

2. Repeat question 1 with the discs replaced by coloured sweets. When a sweet is selected it is eaten and not replaced.

3. The probability that a man selected at random from a population is left-handed is 0.2. The probability that a man selected at random from the same population needs glasses is 0.3.

    (a) Explain why the probability of a man being left-handed or needing glasses is not 0.5.

(b) Draw and label a probability tree diagram to illustrate the above probabilities when a man is chosen at random from the population.

(c) Calculate the probability that the man is right-handed and needs glasses. (SEG Modular)

4. Kate has a box containing 15 chocolates which look identical. Ten have soft centres and the rest have hard centres. She picks a chocolate at random, eats it and then she picks a second chocolate at random and eats it.

(a) Draw the probability tree diagram to represent the outcomes when the two chocolates are eaten.

(b) Calculate the probability that she eats two chocolates of different types.

Kate picks at random and eats a third chocolate.

(c) Calculate the probability that she eats three soft-centred chocolates. (SEG Modular)

5. The probability that Simon passes his driving test at the first attempt is $\frac{1}{3}$. If a test is failed, the probability that Simon passes his driving test at the next attempt is $\frac{7}{12}$. Calculate the probability that Simon passes his driving test at his third attempt. (Hint: draw a tree diagram.)

6. A tube of fruit gums contains 3 red gums, 4 black gums and 3 green gums. Two gums are chosen, at random, and not replaced. What is the probability that they are both the same colour?

Conditional probability

Wendy is looking at a job advert in the paper.

On past experience she estimates that she has a 70% chance of getting the job **if** she can get an interview.

Wendy thinks that she has a two in five chance of getting an interview.

You can write these statements like this:

P (Wendy gets an interview) $= \frac{2}{5} = 0.4$
P (Wendy gets the job **given that** she gets an interview) $= 0.7$

The second of these probabilities is a **conditional probability**.

| A conditional probability is always dependent on events that come before it. |

● Example

Look at Wendy's chances of getting the job advertised in the paper.

(a) Given that she gets an interview, what is the probability that she does not get the job?

(b) What is the probability that Wendy does not get the job?

(a) P (Wendy does not get the job given that she gets an interview) $= 1 - 0.7 = 0.3$

(b) Draw a tree diagram:

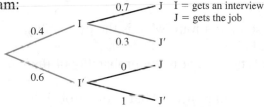

I = gets an interview
J = gets the job

$$P\text{(Wendy does not get the job)} = P\,(I \text{ and } J') + P\,(I')$$
$$= 0.4 \times 0.3 + 0.6$$
$$= 0.72$$

Exercise 7G

1. A man has either egg or toast to eat for breakfast. He drinks either fruit juice or coffee. The probability that he has egg is 0.6. If he has egg the probability that he drinks fruit juice is 0.7 and if he has toast the probability that he drinks fruit juice is 0.2.

    Use a tree diagram to find the probability that he has

    (a) toast and coffee,   (b) coffee,

    (c) coffee given that he has toast.

2. The probability that a person has a disease is 0.02.

    A test for the disease is such that if you have the disease you have a positive test result with a probability of 0.95.

    If you do not have the disease you have a probability of a positive test result of 0.03.

    By using a probability tree diagram or otherwise:

    (a) (i) Calculate the probability that you have the disease and have a positive test result.

    (ii) Calculate the probability that you have a positive test result.

    100 people have a positive result to the test.

    (b) How many of these 100 people would you expect to have the disease?

    (c) Comment on whether the test is worth carrying out.

                                                      (SEG Modular)

3. A fair coin is tossed.

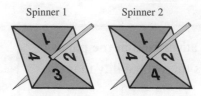

Spinner 1   Spinner 2

If it lands on heads, spinner 1 is chosen.

If it lands on tails, spinner 2 is chosen.

(a) Given that the coin lands heads, what is the probability of the spinner showing a four?

(b) Given that the spinner showed a four, what is the probability that the coin landed on heads?

(c) Given that the coin lands tails, what is the probability of the spinner showing a three?

(d) Given that the spinner showed a three, what is the probability that the coin landed on tails?

Venn diagrams

Venn diagrams provide a means of representing data or probabilities.

Each region contains clearly defined values.

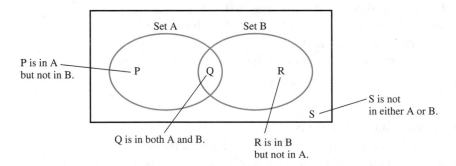

You can use Venn diagrams to solve probability problems.

● **Example**

The probability that a man wears glasses is 0.4.

The probability that a man reads a broadsheet newspaper is 0.7.

The probability that he does not wear glasses and does not read a broadsheet newspaper is 0.1.

Use a Venn diagram to obtain the probability that a man selected at random wears glasses and reads a broadsheet newspaper.

✦ You first draw and label a Venn diagram. (G for the set of men who wears glasses, B for the set of men who read a broadsheet newspaper.)

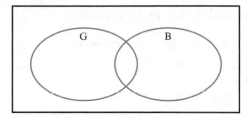

✦ Next you insert the probabilities given.

In this case you can only write in 0.1. This is written in the region outside G and outside B.

The unknown probability for both occurring is represented by a letter (x).

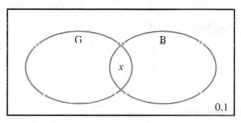

The probability that G occurs is 0.4, so the probability that G occurs but B does not is $0.4 - x$.

Similarly $0.7 - x$ can be entered as shown.

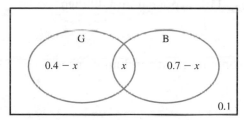

The total probability is 1.

So add them up:

$(0.4 - x) + x + (0.7 - x) + 0.1 = 1$

$1.2 - x = 1$

$x = 0.2$

✦ Now you can write probabilities for each section in the appropriate place.

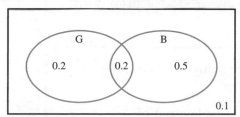

The probability that a man selected at random reads a broadsheet newspaper and wears glasses is 0.2.

Exercise 7H _____

1. The probability that a school pupil is in the football team is 0.4.
   The probability that a pupil is in the cricket team is 0.3.
   The probability that he is in both is 0.1.
   Use a Venn diagram to obtain the probability that he is in neither.

2. All the people at a library are either using the lending library or the
   reference library. The probability that a person uses the lending
   library is 0.9, the probability that they use the reference library is 0.3.
   Draw a Venn diagram to illustrate these probabilities and find the
   probability that a person chosen at random at the library uses both the
   lending and reference sections.

3. The probability that a person watches both EastEnders and
   Coronation Street is 0.6.
   The probability that a person watches neither is 0.05.
   The probability that only Coronation Street is watched is 0.27.

   (a) Draw a Venn diagram to illustrate these probabilities.

   (b) Calculate the probability that a person watches EastEnders.

   (c) Calculate the probability that a person watches either
       EastEnders or Coronation Street but not both.

4. In a group of 200 people, 65 own a dog, 110 own a cat and 40 own
   neither.

   (a) How many own both a dog and a cat?

   (b) What is the probability that a person chosen at random owns a
       dog or a cat but not both?

7.7 Expected frequencies

If you know the chance of something happening then you can work
out how many times it is likely to happen. This is its **expected
frequency**.

> Expected frequency = probability of event happening × number of trials

● **Example** _____

The probability of a particular train service being late is 0.15.

It runs once a day, every day of the year.

(a) How many times would you expect it to be late out of
    30 randomly chosen days?

Over a whole year the train is actually late on 80 occasions.

(b) Is this more or less than what you would actually expect?
(Assume that the year contains 365 days.)

(a) P (late on a particular day) = 0.15
Expected number of late trains out of 30 days = 30 × 0.15
                                              = 4.5

(b) 0.15 × 365 = 54.75
80 is greater than 54.75, so this is a lot more than expected.

> Note: Expected frequencies do not need to be whole numbers.

Exercise 7I

1. A packet of flower seeds contains 80 seeds. The flowers are either red or yellow.
It is known that the probability of a red flower is 0.4.
How many of the flowers would you expect to be red?
What is the probability of a yellow flower?

2. The probability that a non-fiction book is borrowed from the library is 0.2.

   (a) Calculate the probability that a fiction book is borrowed from the library.
   (b) In a week 2500 books are borrowed. How many of these would you expect to be non-fiction?

3. A dice has the numbers 1, 1, 1, 2, 2, 3 on its faces.

   (a) What is the probability of scoring 2?

   The dice is rolled 100 times.

   (b) How many times would you expect to score 2?

4. The probability of a seed germinating is 0.9.
A gardener needs 250 plants.

   (a) How many seeds, in theory, should he sow?
   (b) What is he likely to do in practice?

7.8 Probability distributions

Many uncertain situations fall into distinct categories, and you tend to work out their probabilities in similar ways. These categories are called **probability distributions** and if you recognize them you can work out probabilities very quickly.

The discrete uniform distribution

When you roll a fair dice, each of the six faces is equally likely to occur.

The probability of each number is $\frac{1}{6}$.

You can draw a probability
diagram to show this:

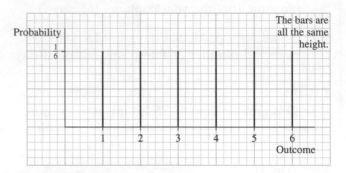

Any distribution of this type is known
as a **discrete uniform distribution**.
In this case the mean and the median
values are both 3.5.

> A discrete uniform distribution occurs when the probability of each and
> every outcome is equal.

If the distribution has outcomes of 1, 2, 3, ..., n then the probability of
each value occurring is $\frac{1}{n}$.

When the values are evenly spaced, you can find the mean and median by:
✦ adding the first and last values, then
✦ dividing by 2.

Example

Random numbers from 0 to 9 are generated using a calculator.
Draw a diagram to illustrate the probability distribution.
(a) What is the probability of each value?
(b) What is the expected mean value?

The probability distribution is a discrete uniform distribution.

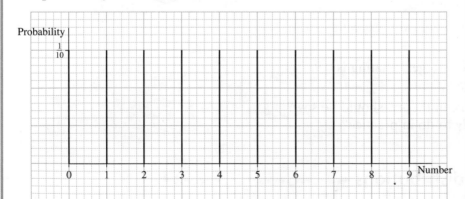

(a) There are 10 values, so the probability is $\frac{1}{10}$.

(b) The expected mean value is $\dfrac{(0+9)}{2} = 4.5$.

Exercise 7J _____

1. Draw a probability distribution for a regular octahedron with the numbers 1, 2, 3, 4, 5, 6, 7 and 8 on its faces.
Write down the mean value.

2. Taxi numbers run consecutively from 1 to n.
Jodie collects the taxi numbers from vehicles she has hired.
The numbers are 27, 35, 64, 62 and 48.

(a) Use the mean to estimate the number of taxis.
(b) Use the median to estimate the number of taxis.

3. Ahmed also records the numbers of vehicles that he has hired.
His data is 23, 42, 10, 6, 89.

(a) Use the mean to estimate the number of taxis.
(b) Use the median to estimate the number of taxis.
(c) Which of these estimates is impossible? Explain your answer.

The binomial distribution

Some events have only two possible outcomes: success or failure, hit or miss, heads or tails. Many of these situations can be described using the binomial distribution.

You can calculate probabilities using the binomial distribution if:

✦ There are just two mutually exclusive outcomes ('success' or 'failure')
✦ The event is repeated a fixed number of times (or 'trials')
✦ There is a fixed probability of 'success'
✦ Each trial is independent of the previous one

Example ─────────────────────────────

A fair coin is tossed four times.

(a) Calculate the probability of obtaining (i) four heads (ii) four tails
(b) One possible outcome could be HTTT.
   (i)   Write out the other ways in which one head and 3 tails could occur.
   (ii) Calculate the probability of HTTT.
   Hence calculate the probability of obtaining
   (iii) one head and three tails
   (iv) three heads and one tail
(c) Copy and complete this table:

Number of heads	0	1	2	3	4
Probability					

(a) (i) $p(\text{one head}) = \frac{1}{2}$

   $p(\text{four heads}) = \frac{1}{2} \times \frac{1}{2} \times \frac{1}{2} \times \frac{1}{2} = \frac{1}{16}$

   (ii) $p(\text{one tail}) = \frac{1}{2}$

   $p(\text{four tails}) = \frac{1}{2} \times \frac{1}{2} \times \frac{1}{2} \times \frac{1}{2} = \frac{1}{16}$

(b) (i)   One way is HTTT.

   The other possibilities are: THTT

   TTHT

   TTTH

   (ii) $p(\text{HTTT}) = \frac{1}{2} \times \frac{1}{2} \times \frac{1}{2} \times \frac{1}{2}$

   $= \frac{1}{16}$

   (iii)   One head and three tails is equivalent to

   HTTT or THTT or TTHT or TTTH

   P(one head
   and three tails) $= 1\frac{1}{16} + \frac{1}{16} + \frac{1}{16} + \frac{1}{16}$

   $= \frac{4}{16}$

   $= \frac{1}{4}$

   (v)   Since heads and tails have the same probability $(\frac{1}{2})$,

   P(three heads and one tail) $= p(\text{one head and three tails})$

   $= \frac{1}{4}$

(c) $p(\text{no heads}) = p(\text{four tails}) = \frac{1}{16}$ (from part a)

   $p(\text{one head}) = \frac{1}{4}$ (from part b)

   $p(\text{three heads}) = \frac{1}{4}$ (from part b)

   $p(\text{four heads}) = \frac{1}{16}$ (from part a)

   The sum of all the probabilities is 1.

   $\frac{1}{16} + \frac{1}{4} + \frac{1}{4} + \frac{1}{16} = \frac{5}{8}$

   So $p(\text{two heads}) = 1 - \frac{5}{8} = \frac{3}{8}$

The table looks like this:

Number of heads	0	1	2	3	4
Probability	$\frac{1}{16}$	$\frac{1}{4}$	$\frac{3}{8}$	$\frac{1}{4}$	$\frac{1}{16}$

Exercise 7K

1. Maya chooses three cards at random, one after the other, from a pack of 52 playing cards.

Red cards are denoted R, and black cards are denoted B.

One possible choice that Maya could make is RBB.

(a) List all the other possible choices that Maya could make.

(b) (i)   Calculate the probability of obtaining RBB.

   (ii)   Calculate the probability of obtaining one red card.

(c) Copy and complete this table.

Number of red cards	0	1	2	3
Probability				

2. Robbie plays darts.
He hits the 'treble 20' once in every three throws.
If he throws three darts, find the probability that Robbie obtains:

(a) (i) one treble 20
  (ii) more than one treble 20

(b) Describe an assumption that you have made.

3. In a multiple choice test each question has three alternative answers with only one correct answer.
Part A of the test consists of four questions. Rukshana guesses the answers randomly.

(a) Explain why the binomial distribution is appropriate in this situation.

(b) Calculate the probability that Rukshana gets all the answers correct.

(c) (i) Calculate the probability that she gets the first answer correct.
  (ii) Hence calculate the probability that she gets exactly one answer correct.

4. A fair dice is thrown three times.

(a) (i) What is the probability of obtaining a square number with a single throw of the dice?
  (ii) What is the probability of obtaining square numbers on all three throws?

> Hint: Square numbers are
> $1 \times 1, 2 \times 2 \ldots$

(b) Calculate the probability of obtaining exactly two square numbers. You may need to list all the outcomes first.

(c) Copy and complete this table:

Number of square numbers				
Probability				

(d) Represent this information on a vertical line diagram

(e) Describe the shape of the distribution.

7.9 Simulation

In a breakfast cereal pack a coloured toy is given away free.

There are 6 different colours of toys.

In order to get the set of six different coloured toys you might be lucky and buy only 6 packets, or you might buy 20 or more packets.

A **simulation** can give an idea of how many packets you would need to buy.

Here are two ways of getting the information.

Method 1

✦ Get all your friends to buy that cereal and see how many packets each bought before they got a set.

Method 2

✦ Match each colour to a number on a dice.

✦ Roll the dice until each number has occurred once:

$$2, 2, 6, 4, 2, 2, 5, 4, 1, 5, 4, 3$$

All six numbers have occurred after 12 packets were purchased.

✦ Repeat the simulation

$$1, 3, 3, 1, 4, 6, 6, 5, 3, 5, 6, 3, 5, 2$$

This time 14 packets are purchased.

✦ Repeat this as many times as you like.

Note: you can use random numbers to generate the data rather than rolling a dice. See page 9 for more information. You can also set up a spreadsheet to select a number between 1 and 6.

> The more times you repeat the experiment, the more confident you can be in the results.

Method 2 is known as a **simulation**.

The advantages of simulation are:

✦ It is quick and cheap.

✦ Using a spreadsheet you can get several hundred pieces of data.

✦ You can adjust the probability to cater for unequal numbers.

Example

At a road junction it is known that 80% of cars turn left and 20% of cars turn right. How would you allocate numbers to simulate cars approaching the junction?

Allocate random numbers as follows:

If you get a 0 or 1 the car turns right.
If you get a 2, 3, 4, 5, 6, 7, 8 or 9 the car turns left.

(Clearly you could allocate any two numbers for the right turn, but you must decide on the allocation before producing the random numbers.)

The following numbers were obtained, using a random number key on a calculator.

5	8	9	1	1	5	2	7	2	2	0	8	6	7	3	2	0	9	3	8
L	L	L	R	R	L	L	L	L	L	L	R	L	L	L	L	L	R	L	L

Exercise 7L

1. On average,   3 months in 10 are wet,
                   6 months in 10 have normal rainfall,
                   1 month in 10 is dry.
   If there are 3 consecutive wet months a particular crop will fail.

   (a) Run a simulation to see how many times in 50 months the crops
       fail due to wetness. (Assume it takes 1 month for the crop to fail.)
   (b) Repeat the simulation.

   For another crop, 2 consecutive dry months cause that crop to fail.

   (c) Use your simulation to determine in a period of 50 months how
       many times this crop would fail.
   (d) Try to explain why on average in every 100 months there would
       only be one occurrence of 2 consecutive dry months.

2. A restaurant offers three choices of roast dinner.
   It is known that 3 out of every 6 customers choose roast chicken,
                    2 out of every 6 customers choose roast lamb,
                    1 out of every 6 customers chooses roast pork.
   20 people are booked in one day.

   (a) Use a dice (or the random number button on your calculator)
       to simulate their choices. Run the simulation several times.
   (b) What is the largest number of selections of roast pork that
       occurs in your simulations?
   (c) What is the largest number of selections of roast chicken that
       occurs in your simulations?
   (d) Explain why the restauranteur may want to run the simulation.

3. A doctor wants to simulate the distribution of boys and girls in
   families with 4 children.

   (a) (i)   Describe how he could simulate families using a coin.
       (ii) What assumptions have you made?
       (iii) Run the simulation using a coin or random number
             generator.

   The doctor has observed that 52% of children are boys.

   (b) (i)   Describe how he could use random numbers to simulate
             families with 4 children.
       (ii) Run the simulation using a random number generator.

4. A statistician analyses football scores on a particular Saturday.
   He records the percentage of teams who score a particular number
   of goals.
   The results are given in the table.

Number of goals	Percent of home teams	Percent of away teams
0	15	30
1	20	30
2	25	20
3	20	15
4	15	5
5	5	0

The statistician allocates random numbers for the home team as follows:

Number of goals	Percent of home teams	Allocation of random numbers
0	15	01 to 15
1	20	16 to 35
2	25	36 to 60
3	20	61 to 80
4	15	81 to 95
5	5	96 to 99 and 00

(a) Use a similar method to allocate numbers for the away team.

From his calculator the statistician obtains two random numbers in the range 00 to 99. They are 89 and 27.

(b) To what result does this selection of random numbers correspond?

Summary

You should now be able to	Check out 7
1 Handle simple probability.	**1** A bag contains 20 red, 30 yellow, 10 green and 40 blue beads. (a) What is the probability that a bead selected at random is (i) red, (ii) not red, (iii) red or green, (iv) red and green? (b) The probability that a hockey team wins its next match is 0.5. The probability that it draws its next match is 0.2. Calculate the probability that it loses its next match.
2 Draw sample space, tree diagrams and Venn diagrams.	**2** Two fair dice are rolled. The two scores are multiplied together. (a) Draw a sample space diagram. (b) What is the probability of a score of (i) 4, (ii) 11? (c) What is the most likely score? A man either travels by car or by bus to work. He is either on time or he is late. (d) Draw a tree diagram to illustrate the possibilities.
3 Calculate combined probabilities.	**3** A man either travels by car or by bus to work. The probability that he travels by car is 0.6. If he travels by car, the probability that he is late is 0.3. If he travels by bus, the probability that he is late is 0.2. Use your tree diagram to calculate the probability that on a randomly chosen morning he is late.
4 Carry out a simulation.	**4** A bus can seat 30 people. The bus is empty when it arrives at the first bus stop. A random number of people between 0 and 10 get on the bus at each stop and a random number of people between 0 and 5 get off at each stop. Use a simulation to find out when some of the people at a bus stop cannot get on the bus.
5 Calculate probabilities using the discrete uniform and binomial distributions.	**5** (a) A fair roulette wheel is numbered from 1 to 36 inclusive. Write down the probability that the next number (i) is less than 8, (ii) is between 12 and 20 inclusive. (iii) What is the mean score when the wheel is spun 100 times? (b) The probability that a woman is late for work is 0.2. (i) Calculate the probability that in a 5-day week she is late at most twice. (ii) What assumption do you need to make in order to calculate the probability?

Revision Exercise 7

1. James conducted an experiment to find out the probability that a
    drawing pin lands point upwards. He threw a drawing pin 300
    times. He recorded the results in this table.

Number of throws	Number of times drawing pin landed point upwards	Relative frequency
First 50	17	0.34
First 100	35	0.35
First 150	54	0.36
First 200	70	
First 250	90	
First 300	111	

    (a) Copy and complete the table.
    (b) Which of these relative frequency results gives the best estimate
        of the probability of a drawing pin landing point upwards?
        Give a reason for your answer.   [NEAB]

2. A pupil threw a coin 40 times and recorded the results on the
    graph below.

    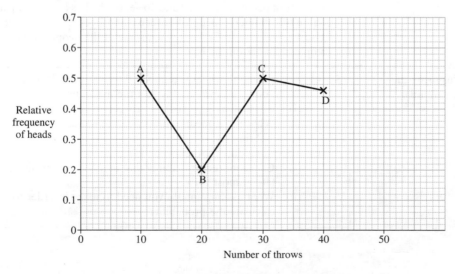

    (a) How many heads were obtained in the first ten throws?
    (b) One of the points was plotted incorrectly. Which point was
        plotted incorrectly? Explain your reasoning.
    (c) The pupil threw the coin another 10 times so that after
        50 throws there were 23 heads. Copy the graph and plot the
        next point.   [NEAB]

3. The pupils in a small school were classified by hair colour and eye colour.

	Brown eyes	Not brown eyes
Fair hair	3	57
Not fair hair	42	38

(a) How many pupils were there in the school?
(b) What is the probability that a pupil chosen at random has brown eyes?
(c) What is the probability that a pupil chosen at random has fair hair?
(d) A pupil with brown eyes is chosen at random.
   What is the probability that the pupil has fair hair?
(e) You are told the eye colour of a pupil from this school.
   You have to guess whether the pupil has fair hair.
   Explain how you would do this.   [NEAB]

4. The sweets in a jar are red, yellow or orange.
The probability that a sweet, chosen at random, will be red is $\frac{1}{4}$ and the probability that it will be yellow is $\frac{2}{5}$.
If I choose one sweet at random, what is the probability

(a) that it will be red or yellow,
(b) that it will be orange,
(c) that it will be white?

There are 60 sweets in the jar.

(d) Calculate the number of red sweets.

(e) $\frac{1}{3}$ of the red sweets have soft centres and the rest have hard centres.
   How many red sweets have hard centres?   [NEAB]

5. In a board game, a counter is moved along the squares by an amount equal to the number thrown on a fair dice.
If you land on a square at the bottom of a ladder you move the counter to the square at the top of that ladder.

18	17	16
13	14	15
12	11	10
7	8	9
6	5	4
Start	1	2	3

(a) What is the probability that a player reaches square 4 with one throw of the dice?

(b) What is the probability that a player can reach square 7 with one throw of the dice?

(c) What is the probability of taking two throws to get to square 2?

(d) List the three possible ways to land on square 18 with exactly three throws of the dice.

(e) Calculate the probability of landing on square 18 with exactly three throws of the dice.   [SEG]

6.

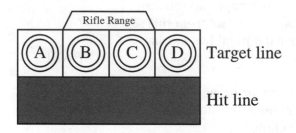

When a letter in the target line is hit, its light goes out and that letter lights up in the hit line. When letters B and D are hit, the letter pair BD lights up in the hit line as shown below.

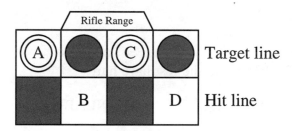

Only two shots are allowed and each shot will **always** hit a different letter.

(a) Write down all the letter pairs that can occur on this rifle range.

For each shot there is an equal probability of hitting any of the remaining letters.
Peter and Francine each take a turn.
With the first shot Peter hits the letter A.

(b) What was the probability of this happening?

(c) Find the probability that Peter's second shot hits letter C.

Francine finishes with the letter pair BD.

(d) What was the probability of getting this result?

7.   SWEET SIXTEEN

START →	1	2	3	
	8	7		5
	9		11	12
	16	15	14	13

'Sweet Sixteen' is a game for any number of players. To play the game, players take it in turns to throw a fair die and then move their counter the number of places shown uppermost on the die. If a player lands on one of the shaded squares the player must start again. The first player to *land on square 16* is the winner. If a player would move past square 16 on a throw, the player is not allowed to move and misses that turn.

(a) What is the probability that a player lands on a shaded square on the first throw?

(b) A player moves to square 3 on the first throw. What is the probability that the player lands on a shaded square on the second throw?

(c) (i) A player is on square 12 after three turns. Write, in the order thrown, three scores the player could have had.

   (ii) In how many different ways could a player have reached square 12 with three throws? Show working to support your answer.

(d) (i) What is the minimum number of turns necessary to complete the game?

   (ii) What is the probability of this happening?   [SEG]

8. The pupils in a class are classified by gender, hair colour and eye colour. The diagram shows that 7 boys have dark hair and brown eyes.

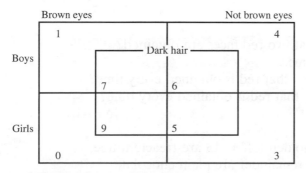

(a) How many pupils are in the class?

(b) What is the probability that a pupil chosen at random has dark hair?

(c) One part of the diagram is labelled 5. What does the diagram tell you about these 5 pupils?

(d) A girl is chosen at random. What is the probability that she has dark hair?

(e) A pupil with dark hair is chosen at random. What is the probability that the pupil does **not** have brown eyes?   [NEAB]

9. Ian buys four hyacinth bulbs. One will produce a pink flower, one will produce a white flower and two will produce blue flowers. He intends planting two bulbs together in a pot.

(a) Draw a tree diagram showing the different ways of choosing his two bulbs.

(b) Use the tree diagram to calculate the probability that the first two flowers will be
   (i)   both blue,
   (ii) flowers of different colours,
   (iii) both pink.   [SEG]

10. A company decides to tender for 3 contracts, A, B and C. The probability that it will obtain contract A is 0.2, contract B is 0.4 and contract C is 0.3. The contracts are awarded independently of one another.
Calculate the probability that it will obtain

(a) no contracts,

(b) only one contract,

(c) at least **two** contracts.   [NEAB]

11. A fair eight-sided dice has 6 red faces and 2 white faces.
The dice is thrown once.

(a) What is the probability of obtaining
   (i) a red face,
   (ii) a white face?

(b) The dice is thrown twice.
   What is the probability that two red faces are obtained?

(c) The dice is thrown four times.
   (i) What is the probability that red is obtained every time?
   (ii) Is it likely or unlikely that red is obtained every time?
                                                        [NEAB]

12. In a bag of 50 sweets, 10 are golden toffee, 13 are treacle toffee, 15 are milk chocolate and the remainder are plain chocolate.
Tim chooses one of the sweets at random.

(a) What is the probability that it is a plain chocolate?
(b) What is the probability that it is a toffee?
(c) Sarah chooses from an identical full bag of sweets. The sweet she takes out is a toffee.
What is the probability that it is a treacle toffee?   [NEAB]

13. A green bag contains 8 one pence coins and 5 two pence coins.
A yellow bag contains 9 one pence coins and 6 two pence coins.
A coin is selected at random from the green bag and placed in the yellow bag.
A second coin is then selected at random from the yellow bag and placed in the green bag.

(a) Draw a probability tree diagram illustrating the two selections.
(b) Calculate the probability that the sum of money in each bag is unchanged after the two transfers.

A third coin is then selected at random from the green bag and placed in the yellow bag.

(c) Calculate the probability that the sum of money in each bag is unchanged after the three transfers.   [SEG]

14. Three women A, B and C, share an office with one telephone. Calls for the office arrive at random during working hours in the ratios $3:2:1$ for A, B, C respectively.
The nature of their work means that the women leave the office independently at random times, so that A is absent for $\frac{1}{5}$ of her working hours and B and C are each absent for $\frac{1}{4}$ of their working hours.

(a) On occasions when the telephone rings during working hours, find the probability that no one is in the office to answer the telephone.
(b) Use the above information to complete a tree diagram.
(c) (i) On occasions when the telephone rings during working hours, find the probability that a caller wishes to speak to C and is able to do so.
  (ii) The phone rings. What is the probability that the office worker requested is available?   [NEAB]

15. Parveen plays Roll-Ball at the fairground. She has to roll six balls, all of which must score. When she has rolled four balls the position is shown in the diagram.
So far she has scored

$$1 + 1 + 2 + 2 = 6$$

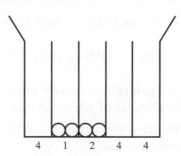

The winning totals are

6   7   8   9   10

Parveen is equally likely to score any number when she rolls a ball.

(a) Which of the winning totals can she still achieve?
(b) How can she achieve these totals?
(c) What is the probability that the next ball scores 1 or 2?
(d) Fill in the missing probabilities on a copy of the tree diagram.

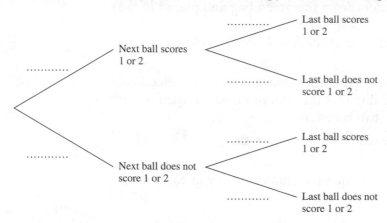

(e) What is the probability that Parveen achieves a winning total?
     Explain your reasoning.   [NEAB]

16. A school debating team has six members, four girls and two boys.
     Two members are to be chosen at random to lead the debate. The
     names of the six members are written on pieces of paper and
     placed in a hat. Two pieces are then chosen at random.

     (a) Draw a tree diagram to illustrate the possible outcomes.
     (b) Find the probability that both girls are chosen.
     (c) Find the probability that at least one boy is chosen.   [NEAB]

17. A gamekeeper wants to estimate the number of pheasant chicks in
     an enclosure.
     He catches and rings 100 chicks and returns them to the enclosure.
     The next day he captures 50 chicks and of these 10 are ringed.

     (a) Estimate the probability that a chick selected at random is
          ringed.
     (b) Use this probability to estimate the number of chicks in the
          enclosure.
     (c) Explain how the gamekeeper could obtain a more accurate
          estimate of the number of pheasant chicks in the enclosure.

     In a different enclosure a gamekeeper catches one chick, notes if it
     is ringed, and returns the chick.

He does this 10 times.

The probability that a chick is ringed is 0.1.

(d) (i) Calculate the probability that 2 chicks are ringed.

    (ii) Calculate the probability that 2 or fewer chicks are ringed.

[SEG]

18. Robin shoots arrows at a target. The probability that he will hit the target with any one shot is 0.6.

(a) Use the binomial distribution to work out the probability that in five shots he will hit the target exactly three times.

(b) Copy and complete the table to show the probability distribution of the number of hits in five shots.

Number of hits in five shots	0	1	2	3	4	5
Probability	0.01	0.077	0.2304			

(c) There are two round in a competition.
In each round there are five shots.
The score is the sum of the number of hits in each round.
Using the table, calculate the probability that Robin will achieve a score of two in the competition.   [SEG]

19. (a) Match one of the following probabilities to each of the events, (i) to (iv), below. Use each probability only once.

$$0, \ \tfrac{1}{4}, \ \tfrac{1}{2}, \ 1$$

(i) A card is taken, at random, out of a pack of 52 playing cards. The probability that it is a red card.

(ii) The probability that, if this month is June, next month is July.

(iii) The probability that it will **not** rain for a year in Britain.

(iv) In a bag of sweets there are equal numbers of red, yellow, green and orange sweets. The probability of taking out a yellow sweet, at random.

(b) In a class of 30 children a survey was carried out to find out how many children liked chocolate and how many liked ice-cream. The diagram shows the results, but the region marked A is not filled in.

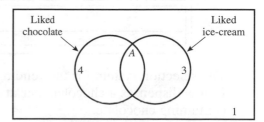

(i) Write the correct number in region A.

(ii) What can you say about the children in region A ?

(iii) If one child is chosen at random, what is the probability that the child liked ice-cream but not chocolate?

(iv) One of the children who liked chocolate is chosen at random. What is the probability that the child also liked ice-cream?

[NEAB]

20. The probability that a person passes their driving test at the first attempt is 0.65. If they fail the first time the probability that they pass at the second attempt is 0.8. If they fail the second time the probability that they pass at the third attempt is 0.85.

(a) Copy and complete the tree diagram to illustrate the probabilities when a person takes the driving test three times.

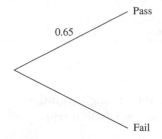

(b) Calculate the probability that a person picked at random

(i)   passed at the second attempt

(ii) passed by the third attempt.   [SEG]

21. A vending machine dispenses plain, mint and orange chocolate bars. The machine is full.

| Plain 1 | Plain 2 | Plain 3 | Plain 4 | Plain 5 | Mint 6 | Orange 7 |

Please remove your selection

HERE

SELECT
1 2 3 4 5 6 7

The selection system for the vending machine is broken.
It now dispenses a chocolate bar at random from any position containing chocolate.

(a) If one bar is dispensed, what is the probability that it is **not** mint chocolate?

(b) If two bars are dispensed, what is the probability that they are **both** plain chocolate?

The machine is used until there is only **one** bar in **each** of the seven sections.

(c) Draw a tree diagram to show the probabilities and outcomes for the next **two** chocolate bars dispensed.
(d) When there is just one bar left in each section, calculate the probability that **no** plain bars will be dispensed in the next two selections.   [SEG]

22. A company produces chocolate.
    60% of its production is milk chocolate and the rest is plain chocolate.
    Of the milk chocolate bars 75% are large and the others are small.
    Of the plain chocolate bars 70% are large and the others are small.

    (a) Draw a clearly labelled tree diagram to represent these probabilities.
    (b) A bar of chocolate is selected at random. Calculate:
        (i)   the probability that it is a small bar of milk chocolate
        (ii) the probability that it is a small bar of chocolate.
    (c) If 270 small bars of chocolate were selected at random, how many of them would you expect to be milk chocolate?   [SEG]

23. The probability that a girl is left-handed is 0.2.
    A group of 10 girls is selected at random.

    (a) Why is the binomial distribution a suitable model for calculating the probability of the number of left-handed girls in the group?
    (b) (i)   Calculate the probability that there are two left-handed girls in the group.
        (ii) Calculate the probability that there is one left-handed girl in the group.
        (iii) Calculate the probability that there are more than two left-handed girls in the group.   [SEG]

24. A packet of geranium seeds contains 12 seeds.
    The probability of a geranium seed germinating is 0.95.
    Use the binomial distribution to calculate the probability, to 4 decimal places, that:

    (a) all 12 seeds germinate,   (b) exactly 11 seeds germinate,
    (c) exactly 10 seeds germinate, (d) 9 seeds or less germinate.
                                                            [SEG]

8 Coursework

> *The purpose of models is not to fit the data,*
> *but to sharpen the questions.*
> Samuel Karlin

Coursework often involves collecting your own data.

This unit will show you how to

+ Choose a topic to study
+ Write a hypothesis
+ Plan the data collection

Before you start

You need to know how to	Check in 8
1 Identify variables. (See Unit 1 page 2)	**1** State (a) a qualitative, (b) a continuous, (c) a discrete variable to do with a pair of spectacles.
2 Design tables (See Unit 1 page 20)	**2** Design a two-way table to collect age and gender of a group of people.

Guidelines for candidates

This is the official advice from AQA:

✦ Make the aim/purpose of the investigation clear and include a reason for choice. State clear hypotheses, for example,

> 'The average number of occupants per car on Long Street varies according to the time of day.'

✦ A set of inter-related hypotheses at higher tier is expected.

✦ State how you are going to investigate the situation. Include:

- the nature of your enquiry (experiment, survey ...);

- the data needed (what data is relevant; ensure that there is enough data);

- the method to be used for collecting the data, giving reasons for choice;

- sampling techniques that are clearly described and sources of primary and secondary data that are acknowledged.

✦ Decide on what exactly needs to be measured, collect the data from suitable sources and consider the best way of recording the data. Where applicable, decide on an appropriate sample size.

✦ Justify choices (e.g. newspapers, books for comparison).

✦ A pilot survey often finds faults in questionnaires. Try out your ideas in order to spot potential pitfalls or fruitful areas and give reasons for changes from the pilot to final survey.

✦ Use computers and calculators in your work whenever you can. Ask your teacher whether there is any relevant software that you can use. Data handling techniques to sort and re-sort data according to different criteria are easier using a computer.

Guidelines for candidates (continued) ▬▬▬▬

+ Keep brief rough notes or a log of any decisions, discoveries, thoughts, observations or ideas which occur.
  Write up your work in stages rather than leaving it all to the end.

+ Represent your data in the most appropriate way that yields the most information and state reasons for the choice. Ensure graphs and diagrams are clearly labelled – some software does not do this well! Ensure a good variety of appropriate diagrams, giving reasons for choice.

+ Choose the best representative value for your data (e.g. the median). Do not be repetitive. Ensure a good range of appropriate techniques, including some higher tier techniques if appropriate.

+ Make sure that you interpret all diagrams and calculations in the context of the original aims or hypotheses.

+ Some of your results may lead you to 'change direction' and investigate a particular aspect further, or to gather more data for comparison.

+ For your conclusion, write a summary of your results and make suggestions for further investigations. Also acknowledge any limitations due to the type of sample possible, availability of data or other factors. Discuss points for possible further investigation.

+ Only make statements or draw conclusions that your work entitles you to do. Your conclusions should link your results and interpretations back to your original aims and hypothesis.

+ Include the titles and authors of any books consulted.

8.1 Specify the problem

The purpose of your coursework is for you to investigate a topic that interests you.

First you should decide on your research question. Your area of research may need to be broken down into sub-questions.

Noah's favourite fruit was apples. He noted that the size and price of the same type of apple varied considerably between shops and he decided to research the best value for money.

What questions could Noah ask in his research?

First he could compare the size of apples against their price.
Then he could compare the wastage (rotten apples) from each shop.
Finally he could assess the taste and quality of the apples.

Noah could then break down his research into three sub-questions:
+ What is the relationship between weight and cost?
+ What is the ratio of good and bad apples for different shops?
+ Which shops sell the best-tasting apples?

Stating a hypothesis

When you have decided on your area of research you should specify a **hypothesis** to be tested.

> A hypothesis is an idea or theory made on the basis of limited evidence. It is a starting point for an investigation.

Your initial ideas may need refining so that the final hypothesis is easy to research.

A hypothesis should:

✦ Be stated clearly.

✦ Be testable.

✦ Be limited in scope.

✦ State the relationship between the variables.

Jenny lives in Hovelton and has two children at Primary school. She wants to research why some children read better than others.

She had an initial hypothesis.

She reflected on her hypothesis.

She revised her hypothesis.

She reflected on it once more.

She now had a well defined hypothesis she could easily test.

But still, the variables 'reading ability' and 'attitude towards education' need to be better defined.

Exercise 8A _____

For the following situations write out a workable hypothesis.
State the variables that will be involved and what data you would collect.
There may be more than one acceptable answer for each situation.

1. Matthew, aged 10, received £10 pocket money each month. His older brother, aged 12, was given £12 pocket money each month.

2. Gemma had to choose subjects to study at A level. To help her decide she compared GCSE results with A level results.

3. Charles Dickens wrote books with long sentences and many adjectives because he was paid for his writing according to the number of words it contained. Authors today are paid a percentage of the book price.

4. Gavin wanted a car for his eighteenth birthday. He became aware that he was always looking at cars. Older cars, he noted, were mostly beige or yellow.

5. Ben's mum was always saying, 'Just a minute' and then taking ages before she came to answer him. He wondered if people actually knew how long a minute really was.

6. Boys were always hogging the computer at school. The boys that seemed to spend a lot of time on the computer and talking about the computer games they played at home were also very good at games that involved catching balls.

7. On a school trip visiting churches, Reuben spent some time looking at the gravestones in the churchyards. He knew that nowadays we are meant to be living longer.

8.2 Plan the data collection

Once you have decided what data to collect, you will need to think about how you are going to collect it.

> The methods that you can use to collect data are explained in Unit 1.

+ If you collect **primary** data, you must keep a record of the process you used to collect the data.
+ If you collect **secondary** data, you need to say where it was published. You also need to say who collected it and why it was collected.
+ If you conduct an **experiment** you need to explain the method you used to collect the data.

> Your research may involve both primary and secondary data or only one of them.

You should justify your choice of method by explaining why it is appropriate to your research and by comparing it with possible alternatives.

Example

Katherine became vegetarian. She was interested in whether more young people are becoming vegetarian and their reasons.

Discuss how she could use both primary and secondary data in her research.

She could:

+ Search the internet, looking at other people's research and then carry out her own survey to see if what she has read is reasonable.
+ Carry out a survey based on personal beliefs and then compare her results with a larger published survey.

Remember: Before you carry out a survey you should first carry out a pilot study to check that your survey is going to give you the information you want.

There is more about pilot studies on page 13.

If you carry out a survey that involves a questionnaire, you should include a copy of your pilot questionnaire as well as your main questionnaire.

You should explain any changes you make to your pilot study.

Exercise 8B

For each of the following hypotheses, discuss how you may use primary and/or secondary data.

There may be more than one acceptable answer for each hypothesis.

1. A hungry mouse will learn the route through a maze quicker than a mouse that is better fed.

2. Nowadays women are choosing to have their first child at a significantly older age than women in the 1950s.

3. Orange Smarties taste different to all other coloured Smarties.

4. More road accidents involving a drunk driver result in a death than accidents in which a driver is not drunk.

5. Asian countries are richer than American countries.

6. If it rains on St. Swithin's Day, it will rain every day for the next forty days.

8.3 Collect the data

p266

p269

If you collect lots of data, you will probably need to summarize it in table form. You should design your table to be clear, concise and easy to use for analysis.

● **Example**

Joshua was conducting an experiment to find out how good people were at estimating. He had two identical lengths of string glued to pieces of card.

One piece of string was glued straight:

One piece was stuck down in a wavy line:

Joshua wanted to research:

1. Is it easier to estimate a length as a straight or a wavy line?
2. Are boys better at estimating than girls?
3. Which age group is best at estimating length?

Design a table to collect all the data that Joshua needs.

Here is a table that Joshua could use:

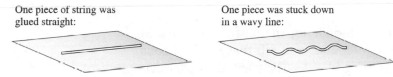

Gender (M/F)	Age	Straight length (cm)	Wavy length (cm)

> Age is personal. Joshua may need to specify an age range to collect this data.

Exercise 8C

For each question design a table that will collect all the data needed for the survey or experiment.

1. Alfie listened to many different types of music. He noticed that the length of the tracks often differed with the type of music he played. Alfie wanted to find out (a) if some categories of music tended to produce longer tracks and (b) whether there was a difference in length of tracks produced now compared with ten years ago.

2. Earl read in a book that height was related to arm span and hand span was related to wrist measurement. He decided to carry out a survey to see if there was any truth in these statements.

3. Freya's mum had become very forgetful as she got older. Freya decided to carry out an experiment to see how many objects placed on a tray people of various ages could remember.

4. Penelope was opening a restaurant. As part of her market research she wanted to find out whether there was any variation in (a) the number of table bookings and (b) numbers per table for different days of the week.

5. Supermarkets sell their own-label products alongside brand-named and economy versions. It is not always possible to detect differences from the ingredients listed. Is it possible to distinguish between the products by taste alone?

6. Claire had been tinting her hair blonde for the past three years. She was aware of the phrase 'dumb blonde', and felt that she had been treated differently since becoming blonde. She wanted to conduct research to see (a) if there was any truth in the phrase and (b) whether or not the phrase was related to age.

Choose **any two** of these questions to suggest ways in which the research can be extended. Extend your table and state why you would want to collect the extra data.

8.4 Represent and analyse the data

Statistics are no substitute for good judgement.
Henry Clay (1777–1852)

By now you should have:
+ Decided on your research question.
+ Formed your hypothesis.
+ Planned your research.
+ Collected your data and organized it.

You can now begin to represent and process your data.

Representing your data

You should illustrate your data with statistical diagrams.

Your choice of diagrams should always be the most suitable for the data.

Analysing your data

To analyse the data fully you will need to select and calculate appropriate measures of location and spread.

> You will need to refer back to Units 3 and 4 for pictorial representations and for summary statistics.

Example

Look back at Joshua's experiment on page 253.

(a) What diagrams could Joshua use to illustrate his data?

(b) What analysis could Joshua perform on his data?

(a) Joshua could illustrate his data with:
- ✦ A scatter diagram of straight length error against age.
- ✦ A scatter diagram of wavy length error against age.
- ✦ Histograms showing the distribution of errors, separated by gender.
- ✦ Box-plots of the data for girls and boys.

(b) To analyse the data, Joshua could:
- ✦ Calculate means and standard deviations of the errors in estimation.
- ✦ Sort his sample by gender and repeat the calculations.
- ✦ Sort his sample into age groups and repeat the calculations.
- ✦ Find Spearman's rank correlation coefficient between the straight and wavy line errors.

> **Note:** There are other representations and calculations that Joshua could use.

> If you have a lot of data then it is sensible to use computers for both graphical work and computations. You will need to select which type of graph to use and which calculations are to be carried out (don't just print off all that the program allows) and give reasons for your choices. See Units 3 and 4 for more information.

8.5 Interpret and conclude

> *A conclusion is the place where you got tired of thinking.*
> Arthur Bloch (Matz's Maxim)

You should **interpret** your tables, diagrams and calculations, referring back to the original hypothesis. Your interpretation should include a detailed explanation of all your findings.

In your **conclusion** you should:

- ✦ Write out a summary of your results.
- ✦ Acknowledge any limitations within your study.
- ✦ Discuss points for possible further investigation.
- ✦ Only make statements within the limitations of the scope of your work.

> It is especially important to interpret and discuss graphical work and computations that have been carried out by a computer.

Further work

Some of your results may inspire you to investigate a particular aspect of your investigation further.

Any further investigation should be interpreted both in relation to a new question and to the original hypothesis.

Summary

The processes involved in statistical research are cyclical in nature:

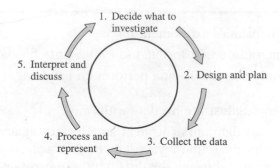

1. Decide what to investigate
2. Design and plan
3. Collect the data
4. Process and represent
5. Interpret and discuss

The area of research chosen to investigate should involve:

> The design and plan of an overall strategy. This should identify aims, hypotheses, data and variables.

> The selection and collection of appropriate data, including a description of the data type and sampling method.

> Processing and representing the data should include
> ✦ making comparisons and contrasts through measures of location and spread,
> ✦ drawing tables and graphs,
> ✦ sorting and resorting the data.

> Interpretation and discussion should relate results, tables and graphs to the original hypothesis.

This may lead to:

> A decision to investigate further.

Summary

You should now be able to	Check out 8
	'... *They measured my right thumb, and desired no more; for a mathematical computation that twice round my thumb is once round the wrist and so on to the neck and waist ...*' Jonathan Swift, *Gulliver's Travels*. Use the extract to ...
1 State a hypothesis	1 Write out a hypothesis.
2 Plan an investigation to collect data	2 State what data you need to collect and how you will collect it.
3 Choose appropriate graphs and calculations	3 Comment on which graphs and calculations you think would be most appropriate.
4 Summarize your results and draw conclusions	4 Say what comparisons you hope to make in your summary.

Coursework ideas: 'These have worked' ▬▬▬▬

These are suggested ideas from AQA.

Surveys

+ Use of local/school facilities: library, tuck shop, hospital, canteen, sports centre, community centre etc
+ Car and traffic surveys
+ Supermarkets, including comparisons
+ Pocket money, earnings, spending habits
+ Opinion surveys, school uniform, music, smoking, eating habits, social issues
+ Personal statistics, use of body-mass index, fitness
+ TV watching habits
+ Homework

Experiments

+ Reaction times (including use of computer software)
+ Throwing and jumping skills
+ Estimation (length, mass etc)
+ Memory tests
+ Puzzle solving

Advertising or packaging claims

+ Mass of chocolate bars, crisp packets etc
+ Number of sweets in a pack
+ Differences between brands

Secondary data (local, national, global)

+ GCSE/KS3 results
+ Attendance statistics
+ Geographical data – GNP, population, mortality rates
+ Economic data
+ Weather data
+ Traffic statistics, accident data
+ Parish records of birth, marriage, death
+ Crime statistics
+ Use of sports data (Olympics, cricket, football etc)

Comparisons

+ Comparisons of characteristics of newspapers
+ Books intended for different age groups written in different centuries

9 Using ICT in Statistics

ICT can be useful in gathering and processing data but not always:

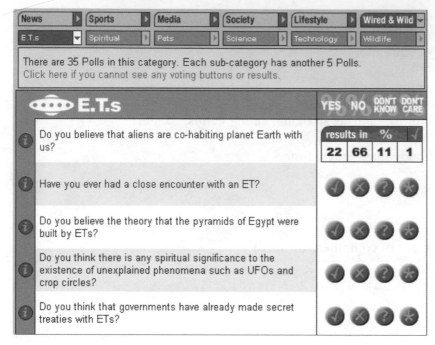

This site:
www.peoples-poll.com
collects much data but the data is not much use as you have no idea of who has contributed.

In this unit you will learn how to:

✦ Use your calculator to perform statistical operations.
✦ Search the internet for Statistical resources.
✦ Produce graphs and perform calculations using Excel 97 spreadsheets.
✦ Run simulations using spreadsheets.

Before you start, you must be able to perform all the calculations and draw all the diagrams on your specification without the use of ICT.

ICT in exams

You are encouraged to use your calculator in an exam. However you must be careful as you will lose all the marks if you get the wrong answer – you would get most of the available marks if you showed the examiner your working.

ICT in coursework

You can use ICT in your coursework but you must show the moderator you understand the method. The best advice is to show all your working for the first use of a calculation and the first use of a graph, and then use ICT to perform the same calculation if you use it again.

Remember:
To gain good marks you must always interpret the statistics you use and ICT will not do that for you!

9.1   Use of a calculator

The GCSE Statistics Scheme of Assessment encourages the use of calculators with statistical functions.

There are two main types of calculator:

✦   One variable calculator and

✦   Two variable calculator

In this section you will see the main statistical operations you can use your calculator for.

The detailed key sequences for each calculator vary considerably, so the unit will encourage you to learn how to use your own calculator to perform the operations. You should have the manufacturer's instruction booklet with you to check what your key sequences should be.

Before data entry

Use the **MODE** or **STAT** function on your calculator to select **Standard deviation** if you are only working with one variable, or select **Regression**, **Linear** if you are working with two variables.

Finding the mean and standard deviation

To find the mean and standard deviation of a set of data you must first work out how to input the data.

You input strings of data one at a time using the data entry key.

You may find the data entry key on the memory key or on the AC key:

Or the data entry key may look like this:

Inputting data from frequency tables is slightly more complex.

✦   Some calculators will allow frequencies to be entered using a multiplication sign:

   ✦ 5×18 is understood by the calculator as the value 5 entered 18 times.

   ✦   Some calculators would read 5×18 as 5 entries of the value 18, not 18 entries of the value 5.

   ✦   Other calculators read 5×18 as the single value of 90. (Then, if the frequencies are small, you can enter the same value repeatedly by pressing the data entry key a number of times – you need to keep a careful count!)

You can check which way your calculator reads the data entry if you have an **n** key. If you press this it will show you the number of entries it is using.

Once you can input the data, the mean and standard deviation are straightforward to find:

+ Press this key for the mean: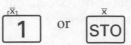

+ Press this key for the standard deviation:

+ Press this for the sum of the values:

+ Press this for the sum of the squared values:

You may also see this key:

This is used at A Level and is not required for GCSE.

Weighted averages

If your calculator allows frequency data entry, you can use it to calculate weighted averages – you just treat the weights as frequencies!

Wrong answers

The most common reason for getting wrong answers is not clearing the memory store before data entry. This is critical – if you start entering data, and are not sure whether you did clear the memory, you may be able to check by asking the calculator for **n** – the number of data values.

Always look at the data to see whether your answer is plausible. If the data values all lie between 4 and 15, the mean can't be 23.73, nor could the standard deviation be 11.4.

Two variable entry

You can also find the mean and standard deviation of pairs of data if you have a two variable calculator.

Make sure you are working in Linear Regression mode.

Some calculators have a straightforward entry sequence:

x, y [DT]

Others are far more complex – check your manual.

Once you can enter your data, you can find the mean and standard deviation of each variable using the keys:

This will help you draw the line of best fit.

Many two variable calculators will also find correlation coefficients for you.

Always refer to your manual and make sure you understand the relevance of the statistic you are using.

Random numbers

You can generate random numbers using the RAND or RAN# key on your calculator.

Most calculators give you three decimal places so you can take samples from up to 1000 pieces of data.

If you only want to sample from a smaller data set, you can fix the number of decimal places displayed:

✦ Find the FIX mode for your calculator and then specify the number of places you want.

Binomial distribution

Some calculators have an nCr key:

which will help you when you are using the binomial distribution with large numbers.

> $^{8}C_{3} = 56$. Check that this is true on your calculator.

You usually have to input the n value, then press the nCr key, then input the r. Check your calculator to make sure this works.

Graphical calculators

A graphical calculator (two variable) will be useful so long as you are prepared to put the effort into learning how to make the best use of it.

Many of them have 'list-based statistics' now, and that is a useful method of entry – what it means is that you can enter (x, y) data into 'lists' for x and y, which allows you to check data entry and so discover mistakes much more easily.

> In most cases you can't print this for your coursework, and you have to draw them by hand for the exam! It will help you check your work though.

You can use them to generate graphs and charts, although Excel is much better for this. A graphical calculator will also draw scatter diagrams and calculate a line of best fit.

Exercise 9A

1. Use your calculator to find the mean and standard deviation of the data sets in Exercise 4A questions 1–3.

2. Use you calculator to find the mean and standard deviation of the data in the Example on page 116.

3. (a) Use your calculator to find the mean of the data set in Exercise 4C question 1.
   (b) If your calculator allows the entry of frequencies, calculate the mean of the data set in Exercise 4C question 2.

> If you use a calculator, there is less need to use coding, though it can help you get a feel for the structure of the data.

4. Find the means of the data sets in questions 1–3 of Exercise 4E.

5. (a) Use your calculator to find the mean and standard deviation of the data sets in Exercise 4N questions 6 and 7.

   (b) If your calculator allows the entry of frequencies, calculate the mean and standard deviation of the data set in Exercise 4N question 5.

6. Use your calculator to confirm your answers to Exercise 7K on the binomial distribution.

7. Use the random number key on your calculator to choose a sample of:

   (a) 34 from 1000 people   (b) 20 from 100 people.

Question 8 is for two variable calculators only.

8. Use your calculator to find the means of the x and y values in Exercise 6C questions 1 to 3.

   Use the values to find the equation of the line of best fit and a measure of correlation. Interpret your results.

> You will find the answers in Exercise 6F question 5.

9.2 Use of the Internet

The Internet is a vast resource available, mostly free of charge, to everyone, but its very size can be frustrating – how do you find the information you want, and in a useable form?

✦ If your search is too specific, and doesn't use exactly the keywords the site creators thought of, then you may not turn up relevant sites.

✦ If your search is too general you can turn up literally millions of sites, and you can't find the most relevant ones.

Since the Internet is literally growing and changing day by day, there is no guarantee that a site that is here today, will still be here tomorrow.

> You will find lots of data which you might like to work with, which will take a lot of effort to get into a form you can use.

For that reason, this section will give some principles to guide you in using the Internet, as well as identifying the 'best sites' currently available.

Really, the opportunities are endless, but it will help if you have a good idea of what information you want before you start.

The information here is given in four categories:

1. Established organizations
2. Data warehouses
3. Specialized sites
4. Miscellaneous

1 Established organizations

If someone is doing the hard work of periodically updating information in a field you are interested in, it can save you a huge amount of time, and frustrated effort. Here are some examples which are likely to exist for a long time:

If the actual link is not still live you can always start at the home site.

+ The Learning and Teaching Support Network has a links page which includes data sources. Go to
  http://www.stats.gla.ac.uk/ltsn/links_stats.html

+ National Grid for Learning: Sidney Tyrrell at Coventry University has a resource page as part of the NGfL. Go to
  http://www.mis.coventry.ac.uk/~styrrell/resource.htm

+ Autograph, based at Oundle School. Go to
  http://www.argonet.co.uk/oundlesch/mlink.html

2 Data warehouses

Data warehouses collect a lot of information on related matters together in one place. Here are some examples:

+ The World Bank has a lot of data which is relevant to a number of curriculum study areas which use data, such as Geography, Economics, Sociology etc. The country tables provide data on more than 50 indicators for 206 of the world's countries. Other links take you to multiple data sets relating to issues such as AIDS and climate change. A good starting point for this currently is
  http://www.worldbank.org/html/schools/data.htm

+ Unicef have another large collection of data, specifically about women and children in countries around the world. Go to
  http://www.unicef.org/statis/

+ The Broadcasters' Audience Research Board stores data on TV viewing figures, which are available on a monthly and weekly basis, and available on a national or a regional basis. Go to
  http://www.barb.co.uk/
  Currently you have to 'register' by entering some contact details, but there is no fee for accessing the data on their site.

BARB do have much more information which they will sell to businesses wishing to analyse viewing habits in order to target advertising efficiently, or schedule programming.

+ The Journal of Statistics Education is an online journal which was published free for a number of years, though it is now subscription based. The archives are still available freely and one of its regular features was data sets, which come with explanations of the context. Go to **http://www.amstat.org/publications/jse/**

+ The Data and Story Library which can be found at **http://lib.stat.cmu.edu/DASL/** can be searched for data in particular subject or topic areas.

3 Specialized sites

These sites give information about a specific subject, or a particular company.

+ The Automobile Association. Go to **http://www.theaa.com/** for information about all sorts of things related to motoring – you can access information about prices of second hand cars, insurance and so on.

+ The National Lottery. Go to **http://lottery.merseyworld.com/** to get information about the winning numbers, the number of jackpot winners in each draw, the ticket sales etc. – all sorts of related data sets which provide opportunities for statistical investigations.

+ Polling and market research organizations are in the data business, so as you might expect their web sites offer a lot of data! Go to **http://www.gallup.com/** or **http://www.mori.com/** or **http://www.nopres.co.uk** for archived data on surveys on just about any topic.

+ The Government. Go to **http://www.statistics.gov.uk/themes/default.asp** This gives you access to a lot of national and regional statistics on themes such as crime and justice, health and care, the economy, education, transport and many others. There are data sets here in csv format, which you can save on disk.

The screenshot shows what the Excel sheet looks like when it opens.

> *To open a csv file in Excel:*
> In the **Type of files** menu, select **All Files** in the folder you have saved the file into, and then select the file.

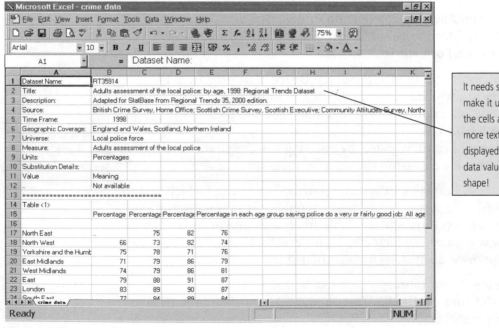

> It needs some work to make it useable – many of the cells at the top contain more text than can be displayed, but the actual data values are in good shape!

✦ The Schools Census. Go to
http://censusatschool.ntu.ac.uk/default.asp and a screen like this
should appear:

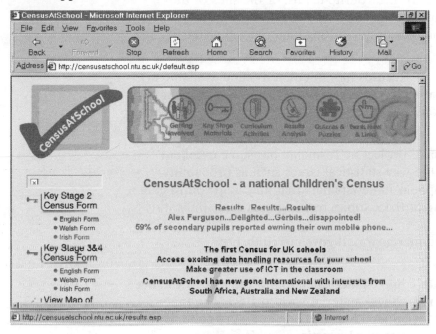

As the screen will be updated periodically, it may not look exactly like this.

This is a site with a lot of information collected about school pupils
in the United Kingdom, linked to the (adult) National Census in
April 2001. It has data sets available for download already in Excel.

Click on the **Results, Analysis** button which brings up different options including **Spreadsheets / Graphs**, and then various files are available to download.

Downloading the **Town** Excel file gives you access to this information:

	A	B	C	D	E	F	G	H	I
1	Key Stage	Gender	DOB	Yr Group	Height	Foot length	Fball team	M Phone	Computer
2	3	M	18/04/85	11	171	27	53	1	1
3	3	M	04/08/88	8	155	23	49	0	1
4	2	M	31/08/91	5	140	17.5		0	0
5	2	F	03/04/92	4	134	20		1	1
6	3	M	14/01/87	9	173	26	41	0	1
7	2	F	19/12/92	4	132	20		0	1
8	3	F	06/08/89	7	149	24	41	0	0
9	2	F	19/08/91	5	126	20		0	1
10	3	F	13/03/86	10	157	23	42	1	1
11	2	M	10/10/90	5	149	25		0	1
12	2	F	12/09/91	4	141	19		0	1
13	2	F	21/02/91	5	130	20		0	1
14	3	M	30/11/88	7	153	23	1	1	0
15	2	F	20/01/93	3	120	18		0	1
16	2	M	23/08/90	6	152	24		0	1
17	3	F	12/09/84	11	171	25	69	1	1
18	3	M	09/01/88	8	157	23	0	0	1
19	3	M	30/03/89	7	165	31	53	0	1

There are 35 columns of data, on 1200 different children here, so this represents a data set which is much larger than anything you would create yourself.

You will use the data in section 9.3 on sorting and searching, and creating summary tables.

4 Miscellaneous

You can search the web for hobbies and sports clubs – use a search engine to find the sites of your favourite teams, or the sites of the national organization for your favourite sport, and explore the links available. Match programmes or advertising material will also carry the web address.

✦ The Guinness Book of Records has an online version available at **http://www.guinnessworldrecords.com/home.asp** where you can get statistical information about the smallest, biggest, fastest everything in the world.

✦ The 'Chance and Data' project in Tasmania has some excellent material looking at the way statistical issues such as probability, graphing, risk etc. appear in newspapers that may help you to gain a better understanding of these topics by looking at real applications of them. Go to **http://www.ni.com.au/mercury/mathguys/mercindx.htm**

✦ Another similar site is at **http://www.dartmouth.edu/~chance/** then follow the link to **Chance News**, which contains archived material back to 1992. The material here tends to be US-orientated but the content is very good, if you are interested in bigger statistical questions and issues.

✦ This site takes you through setting up your own opinion poll step by step: **http:/www.opinionpower.com**

9.3   Spreadsheets

Using Excel 97

Spreadsheets allow you to store and analyse sets of data, perform calculations and construct graphs or 'charts' of your data, and produce reports.

If you use a spreadsheet well it is a very powerful tool.

This unit shows you how to use the basic functions of Microsoft Excel (97) because it is the most commonly available spreadsheet.

> **Remember:** you can use spreadsheets in your Coursework but you must always show the examiner that you understand what you are doing.

Getting started
The basics
A spreadsheet is an array of rows and columns.

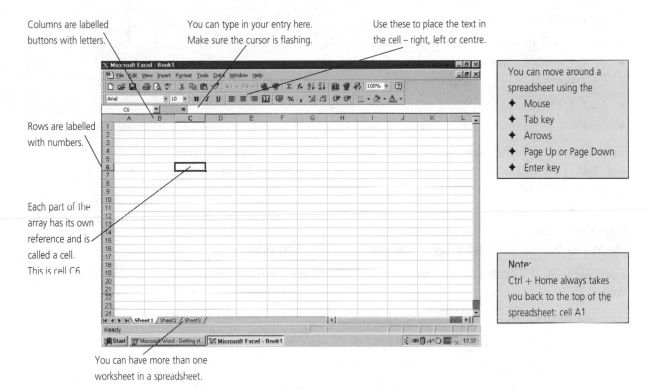

Columns are labelled buttons with letters.

You can type in your entry here. Make sure the cursor is flashing.

Use these to place the text in the cell – right, left or centre.

Rows are labelled with numbers.

Each part of the array has its own reference and is called a cell. This is cell C6

You can move around a spreadsheet using the

♦ Mouse
♦ Tab key
♦ Arrows
♦ Page Up or Page Down
♦ Enter key

Note:
Ctrl + Home always takes you back to the top of the spreadsheet: cell A1.

You can have more than one worksheet in a spreadsheet.

You can change the width of any column:

♦ Put the cursor in the column and then choose **Format** then **Column Width** from the drop down menu, or point the cursor at the boundary of the column header and **click and drag** it to the required width.

Entering text and functions

You can enter **text** in a cell or you can use a function.

♦ Text is any combination of numbers and letters. For example: 10AA109, 20xy, Thursday, 12-284 will all be treated as text by Excel.

You can format any cell to read text as a date or as a money amount (currency) by placing the cursor in the cell and choosing **Format** then **Cells** from the drop down menu.

You can also specify the number of decimal places to be allowed.

You can format a whole row or column or array by highlighting the whole range of cells you want to format:

♦ Hold down the **Shift** key and use the **Arrow** keys to highlight the cells.

You enter a **formula** if you want the spreadsheet to calculate something automatically for you.

♦ You always start a formula with =. The '=' tells Excel it is a formula.

Here are the basic formulae you use:

+ $= C1+C2$ means the sum of cells C1 and C2
+ $= C1 - C2$ means cell C1 minus cell C2
+ $= C1*C2$ means you want to multiply C1 by C2
+ $= C1/C2$ means you want to divide C1 by C2
+ $= C1^{\wedge}C2$ means C1 raised to the power C2.

There are many other formulae that do calculations such as finding the average. You can find a whole list of functions by choosing **Insert** then **Formula** from the drop down menu.

> **Hint**
> =sum(C1:C4) means the sum of cells C1 to C4 and is a shorthand way of adding lots of cells.

Copying formulae

Once you have a formula in one cell, you can copy it to other cells.

For example, if in cell H4 you have written a formula for the total of row 4:

$= sum(A4:G4)$

you can copy this to find the total of rows 5, 6 and 7 by:

+ Highlighting cell H4.
+ Then choose **Edit** then **Copy** from the pull down menu.
+ Highlight the cells H5, H6 and H7 using the **Shift** key and the **Arrow** keys.
+ Press **Enter**. The formula should be copied across.

Example

This is a basic Excel spreadsheet which calculates wages for employees of a company.

This is the title of the worksheet.

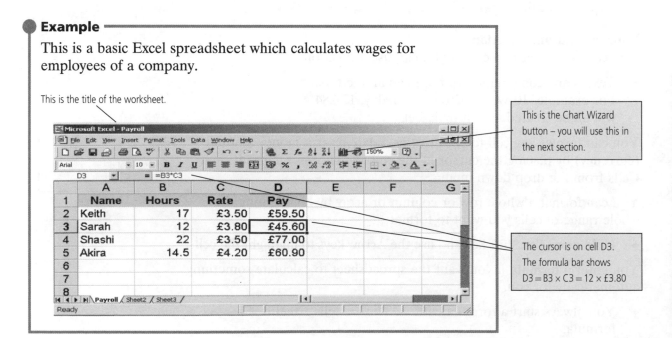

This is the Chart Wizard button – you will use this in the next section.

The cursor is on cell D3. The formula bar shows D3 = B3 × C3 = 12 × £3.80

Exercise 9B _____

1. Reproduce the worksheet in the Example.
   Use formulae to:

   (a) Enter the total number of hours worked in B6.

   (b) Enter the total pay in D6.

   (c) Enter the average hourly rate into F7 (=D6/B6).

   (d) Why is this average hourly rate (£3.71) different to the average of the hourly rates of the four employees (£3.75)?

   > Remember to format the cells as currency.

2. Set up a spreadsheet to calculate the mean of the data set in the Example in section 4.2 on page 92.

3. Set up a spreadsheet to show the table of information in the Example in section 1.10 on page 20. Use the sum facility to extend the table to show both row sums and column sums.

4. Set up a spreadsheet which performs the calculations required for the standard deviation of a frequency distribution. Use the Example at the bottom of page 117.

5. Set up spreadsheets to calculate the weighted averages in Exercise 5A on page 143.

Creating charts

You can use Excel to draw graphs of your data.

Before you even start to do this you must know what type of graph is appropriate to display your data – inappropriate graphs will lose you marks!

Once you know what kind of graph you want to use, you need to highlight the appropriate data on your spreadsheet, using the **Shift** key and the **Arrow** keys.

> To highlight columns or rows that are not next to each other, hold down the **Ctrl** key then move to the required row or column.

For example, in the Example on page 268 you may want to draw a pie chart to show the number of hours worked by each employee. You would need to highlight the name and hours columns.

Now you can start the Chart Wizard by choosing **Insert** then **Chart** from the drop down menu. This is what you will see:

> Or click on the Chart Wizard symbol on the format bar.

1. Choose a chart from the menu.

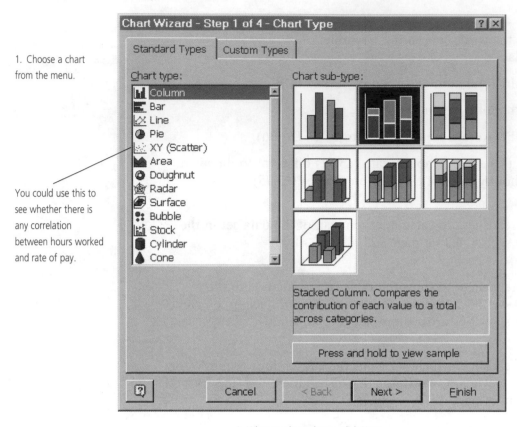

You could use this to see whether there is any correlation between hours worked and rate of pay.

2. Highlight the chart you want to use and then you can see a preview.

3. The terminology is different to your GCSE – make sure *you* use the language from your syllabus.

4. When you have chosen, click **Next>**

Screen 2 of the Chart Wizard gives you a preview of the chart you have chosen.

Screen 3 of the Chart Wizard allows you to customize your chart. You can label it and choose how it should look on the page.

> You can draw a time series chart – look under the **Axes** menu on screen 3.

Screen 4 of the Chart Wizard gives you a range of choices about where you put your chart, and how it will appear.

You can position it on the worksheet the data is in, or in another worksheet. When you update the figures, any charts using those figures will be updated.

You can put it into a Word document – the best way to do this is:

✦ Copy the chart – right click on it to bring up the menu.

✦ Choose **Paste Special** from the **Edit** menu in Word – insert it as an Excel chart or as a picture.

> If you have Paste link enabled then any changes to the spreadsheet should automatically update in your Word document.

If you just cut and paste the chart Word often loses the formats.

Exercise 9C _____

Choose a table from each of the Exercises 3A, 3B, 3C and 3D.

For each table:

Reproduce the table in Excel.

Use the Chart Wizard to draw the chart required.

Insert the chart into Word.

Change some of the data values in the spreadsheet and check that the changes transfer into your chart in Word.

Large data sets

When you have a lot of information, it will be sensible to start to use 'Names' for cells so that it is easier to follow the relationships between the different parts of the sheet.

To use labels in formulae, click **Options** on the **Tools** menu, and then click the **Calculation** tab. Under **Workbook options**, select the **Accept labels in formulae** check box.

The employer from the Example on page 268 is considering awarding a pay rise to his staff.

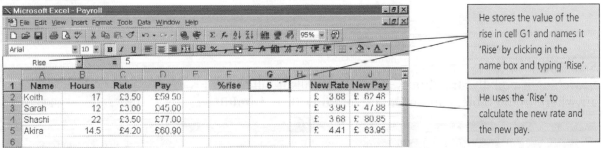

He stores the value of the rise in cell G1 and names it 'Rise' by clicking in the name box and typing 'Rise'.

He uses the 'Rise' to calculate the new rate and the new pay.

The manager can now change the value in G1 to 6, and see how much a 6% rise would pay each of the workers:

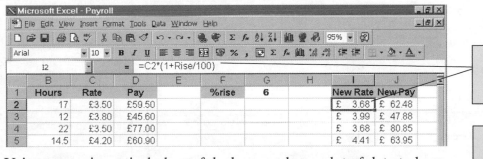

Note that instead of the cell reference G1, the formula uses the name 'Rise'.

Using names is particularly useful when you have a lot of data to keep track of because you can use a description of what you are using rather than a cell reference that may change as you update your sheet.

You can also use Names to define a range of cells rather than a single cell. Details can be found in the Help Index.

Exercise _____

Try to reproduce the worksheet from the Example on page 268.

Try changing the 'rise' and see whether your values change.

Sorting and searching

When you want to use large data sets it is useful to know how to search them and sort them according to the criteria you are interested in.

Sorting

If you are anywhere within the table, and choose **Data** then **Sort**, a dialog box appears giving a choice of criteria for sorting:

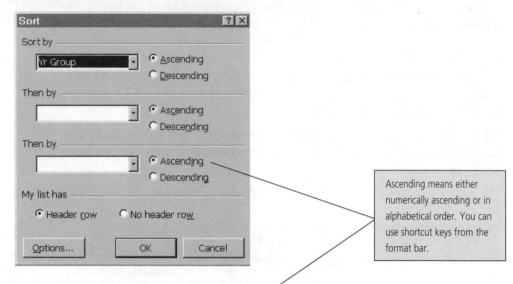

Ascending means either numerically ascending or in alphabetical order. You can use shortcut keys from the format bar.

Searching

Searching is equally straightforward – Excel calls it 'filtering'. From anywhere in the table, choose **Data** then **Filter** then **AutoFilter**.

The drop down arrows contain each of the different entries in the column.

Note: The **'custom'** option available in each drop-down menu allows you to put in more sophisticated criteria for individual columns – **is greater than, does not start with** …

If you select one of those entries the drop-down arrow will turn blue to show that a filter criteria has been applied.

The sheet will show only the rows that satisfy the criteria – the references are from the original table.

Creating summary tables for data

Excel has a built in function called a 'Pivot Table Report' which allows you to create summary tables displaying counts, sums, percentages, averages etc.

The sheet shows some data downloaded from the Schools Census website (see page 265).

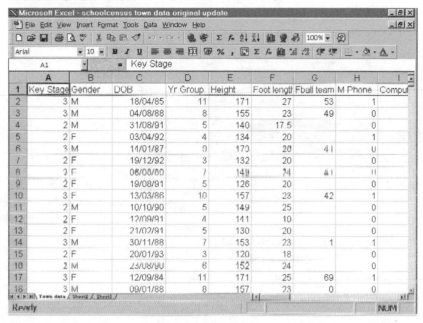

The data set has 1200 'records' i.e. information on 1200 different school pupils, at Key Stages 2 and 3, in town schools. It contains 35 pieces of information for each pupil – the column headers indicate what each field is. The data is coded, as explained on the website. Note that coding for yes/no is easy to interpret.

This is a large data set – too large to understand all at once. You want to pick out particular pieces of information:

✦ Put your cursor anywhere in the table of information, and choose **Data** then **PivotTable Report**.

Step 3 gives the options for what you want the report to contain:

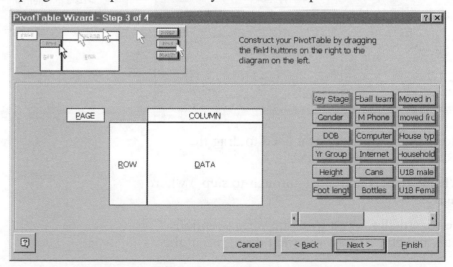

Excel will 'guess' the data range as the whole table. If you want to restrict the range used in the report, you can enter the data range yourself at step 2 of the process.

This is how you would construct a very simple summary table showing how many pupils

✦ own mobile phones,

✦ have computer access and

✦ have internet access at home,

for the different key stages separately:

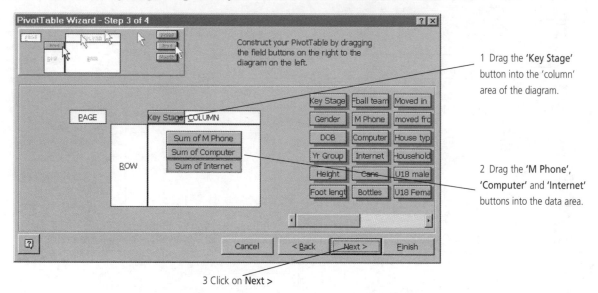

1 Drag the **'Key Stage'** button into the 'column' area of the diagram.

2 Drag the **'M Phone'**, **'Computer'** and **'Internet'** buttons into the data area.

3 Click on **Next >**

You can now choose where you want the summary table to appear – you can opt to place it somewhere on the same sheet as the table of data, but it is often easier to choose to have it in a new sheet.

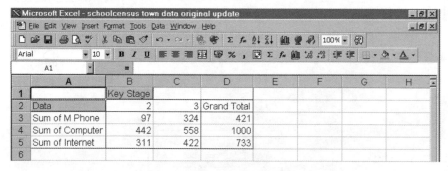

Remember that if the original data entry is amended, the summary table will not automatically update. To update it, choose **Data** then **Refresh data**.

To separate out this information by gender, you need to drag the Gender field into the 'Row' area of the diagram:

Choose **Data** and **PivotTable Report** and move through to step 3 where you can make this change.

Step 4 of the PivotTable wizard allows you to set up various options, including which sub-totals and grand totals you want displayed in the summary table:

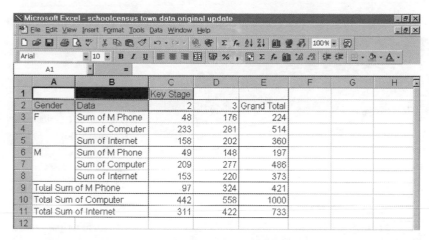

If you go to cell B6, and type 'mobile phone' instead of 'Sum of M phone', the other cells (B3 and D9) which used 'Sum of M phone' are automatically updated:

You can use the PivotTable to find the average of a set of data:
Since 1 was used for access, and 0 when the pupil does not have access, the 'average score' will represent the proportion of pupils in that group who have access. You will need to format the cells to show the proportion as a percentage.

At step 3, choose **Data** and **PivotTable Report** then double click on each of the buttons in the data area of the diagram.

You will see this options box, which allows you to specify that you want the average, and you can format the cells as percentages by clicking on Number.

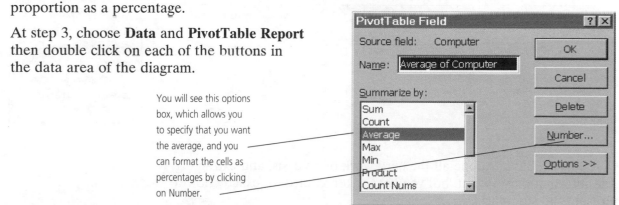

This is the end result:

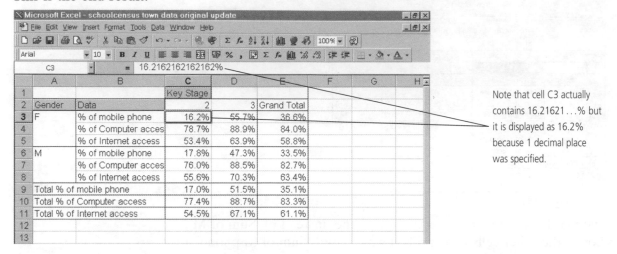

Note that cell C3 actually contains 16.21621…% but it is displayed as 16.2% because 1 decimal place was specified.

Producing charts from PivotTables

You can use the PivotTable to produce a chart.

If you have multiple fields in the Row, Column and Data areas you may find it difficult to format the graph as you want.

You can take a copy of the table, and use **Edit**, **Paste Special** and check the **Values** radio button. You can then rearrange cell orders, combining the multiple field designations into single cells rather than in two columns.

For example, in the above table, changing the data descriptor in row 3 also changes the descriptors in rows 6 and 9. In the copy of the table values you can have row 3 as mobile phone (female) and row 4 as mobile phone (male).

If you intend to use secondary data, it is a good idea to experiment with making charts.

The automated labels in the charts will now be more meaningful.
A comparative bar (column) chart of this table will show pairs of bars for KS2 and KS3, but the female and male information for each technology will be shown as neighbouring pairs of bars.

The PivotTable has done all the hard work of analysis, and extracting the information we want, but you have taken control of exactly how to present the summary.

9.4 Simulations

The use of simulation is becoming an increasingly powerful tool in the world of business. In the past, when an engineer was designing a new aeroplane or car, they would build a physical model first and test its performance in wind tunnels.

Nowadays, the initial testing is done on computers by the use of simulation programs. The same technology lies behind the 'virtual reality' worlds of:

✦ computer video games,

✦ training for fire-fighters in how to deal with certain types of situations where training using a real fire would be too dangerous,

✦ training surgeons in delicate life-saving operations so they do not learn 'on the job' in the way they did previously.

In building a simulation, it is important to realize that you start off simply and then you can make it more realistic in stages – so that it becomes more like the 'real situation' you want to represent.

In this unit you will look at two examples of simulations, so you can see how powerful they can be, and in the process you will see a number of the useful tools spreadsheets have such as random number generators and conditional statements.

Example 1: Dice throwing

Investigate the length of runs observed when throwing a fair dice – how often do you repeat the last number?

How often do you get three in a row the same?

Four in a row?

A spreadsheet will allow the electronic generation of a large set of results in very little time, and because the results are already in a spreadsheet the analysis can also be done very quickly.

If you want to set up an electronic simulation, think through what would you have to do in a 'physical experiment'. Better still, actually do a small-scale version of it so you understand what the key processes are.

These are the results of rolling a dice a number of times:

5	4	1	3	3	4	6	2	1	5	4	5	5	5	2	4	1	6	3	2

Now you can see what is involved in producing the information asked for in the investigation. There are:

+ 0 runs of 4 or more,
+ 1 set of 3 in a row, and
+ 1 pair – or should that be three pairs?

This gives a clue to a process for counting:

+ Compare each number with the one before and if they are the same, then count it as a 'pair'.
+ For three in a row, compare each number with both the two previous numbers, and if the same then count a 'run of (at least) 3',

and so on.

This is the sort of process which computers are extremely good at.

It needs a couple of functions to do this:

1   Generating random numbers between 1 and 6 – to simulate what happens when you throw a dice.

2   Comparing two or more values, and keeping count when they are the same.

To simulate the throwing of a dice, you use two of the math & trig functions:

RAND gives a random number between 0 and 1. To get the six integers from 1 to 6, with equal probability, you

+ multiply this by 6 (giving a random number between 0 and 6)
+ add 1 (giving a random number between 1 and 7 but always less than 7) and then
+ take only the integer part of that.

This may seem complicated, but if you enter:

   $=INT(1+6*RAND())$

in a cell, and then press **F9** repeatedly, you will see a simulation of a sequence of dice throws.

If you enter this into one cell, you can then generate as many throws of the dice as you want by copying the function.

You need to use the logical functions in order to do the tallies of how many repeats, runs of 3, 4, 5 … were observed. The simplest way to do this is to get the computer to return a value of 1 when you see what you want to tally, and 0 when you don't. Then a simple sum will give the total number of occurrences.

5, 5, 5, could be seen as two pairs as well as a run of three.

In the menu of available functions – choose **Insert** then **Function** – the 'statistical', 'logical' and 'math & trig' groups are the ones where most of the useful functions are listed.

Note: You must have the (()) at the end: the () goes with the **RAND** function. Then the last) closes the bracket defining the **INT** function.

In the menu of functions, choose **Logical** then **If**:

This button will also bring up the function dialog box.

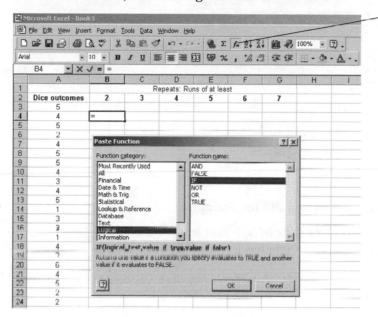

In the dialog box:
1 Enter A4 = A3 [either by clicking on cell A4, typing =, clicking on cell A3 or by typing] in the first box.
2 Enter 1 and 0 in the next two boxes, which are the instructions for what the computer should display when the condition is satisfied, and when it is not.

This function can now be copied into the cells below, as far as you went in generating outcomes.

To check for runs of at least 3 the same, the values need to satisfy two conditions at the same time:

A5 = A4 and also A4 = A3.

If so then A5 represents a run of at least 3.

To do this, you use the nesting capability of the logical statements:

Select cell C5, and go to the **Logical**, **If** statement again.

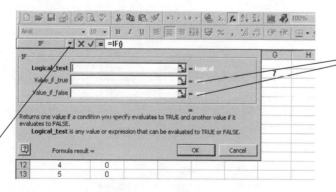

Put in the values of 1 and 0 in the second and third boxes now, so the computer knows what output you want when the criteria have been examined.

The down arrow beside the formula bar allows you to put another logical statement in the **Logical test** section. From the drop down menu, choose **And**. You should see a dialog box like this:

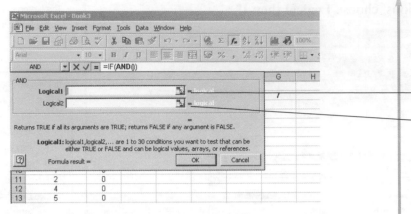

You can enter A5 = A4 in the first box, and A4 = A3 in the second.

You need to copy this function down as you did for the pairs.

You can now use the same idea to count runs of 4, 5 and so on – all that changes is the number of comparisons you list in the **AND** statement.

What remains to be done is to count the total number of occurrences of each length of run.

By using the coding of 1 and 0, the sum of each column gives what you want, or you might want to see it as a proportion.

If your data went down to row 2502, the percentages row would be row 2506. Then cell D2506=AVERAGE(D6:D2502), with the ranges starting at B4, C5, D6 etc. but all ending at 2502.

The table shows the bottom of this simulation sheet – the first column shows the last 5 throws of the 2500 in the simulation. Because there are no repeats in this small block, the other columns are filled with zeros.

The totals and percentages for your version will be different from this, and each time you press F9, you can generate another complete sample of 2500 dice throws, complete with counts and/or percentages.

2	0	0	0	0	0	0
5	0	0	0	0	0	0
1	0	0	0	0	0	0
4	0	0	0	0	0	0
3	0	0	0	0	0	0
	Two	Three	Four	Five	Six	Seven
Totals	410	69	11	1	0	0
	Two	Three	Four	Five	Six	Seven
Percentages	16.4%	2.8%	0.4%	0.0%	0.0%	0.0%

By highlighting two of the rows at the bottom, you can construct a chart to see the visual representation.

Because of the relative size of the numbers of runs of two and of five, six or seven, it is a good idea here to show the data values on the chart.

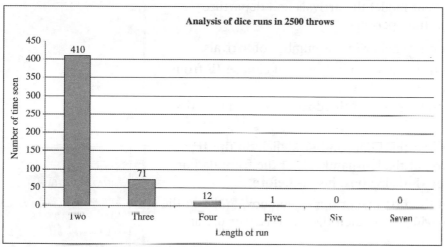

Once this is done, the chart will automatically update, showing the graph of each new set of 2500 throws.

Example 2: Queuing

How a queue behaves in the long run depends on both the rate of arrival at the queue, and how long 'service' takes.

You can model this using a simulation.

✦ New length = old length + arrivals − departures

with the condition that the length can never be negative.

Consider a simple model, where the number of arrivals and departures behave according to this probability distribution:

	0	1	2	3	4
Arrivals	0.2	0.25	0.3	0.15	0.1
Departures	0.05	0.4	0.25	0.2	0.1

> If there was 1 person waiting at the start of the time period, and 3 arrive, but the 'service' random generator says 6 people are served, the queue length at the start of the next period is zero rather than −2!

> There is a probability of 0.1 that 4 people could be served (provided there are 4 people in the queue).

Use the **RAND** function, with these **IF** statements to generate the number of arrivals in a time period:

✦ If the value is < 0.2 you want to return the value 0.
✦ If it is not, but it is < 0.45 then you want the value 1.
✦ Between 0.45 and 0.75 you want the value 2.
✦ Between 0.75 and 0.9 you want the value 3.
✦ Otherwise (over 0.9) you want the value 4.

> These values are the cumulative probabilities of the arrivals:
> $0.2 + 0.25 \, (+0.3) + \ldots$

You can nest **IF** functions to do this. Follow these steps:

1. Enter column headings. The first and second columns will contain the random numbers that model the arrivals and departures. Each row represents a time period.

2. Convert the first random number into a number of arrivals:
   ✦ In cell D2, click on the function button, and choose **IF** from the **Logical** list.
   ✦ Enter **A2<0.2** in the top line of the dialog box, and 0 in the second.

3. Position the cursor in the third line, and now click on the **IF** button which should be at the left-hand end of the formula bar. This opens up another **IF** dialog box, just as before.
   ✦ Enter **A2<0.45** in the first line of this dialog box, and **1** in the second line – this will give 1 when the value is over 0.2 and less than 0.45.

> In the formula bar now, you should see =IF(A2<0.2,0,IF()). Excel will only look at the third line if it has not got a 'true' answer to the condition.

4. With the cursor in the third line, click on the **IF** button in the formula bar again.
   ✦ Enter **A2<0.75** in the first line and **2** in the second and again click on the **IF** button with the cursor in the third line.
   ✦ Enter **A2<0.9** in the first line and **3** in the second, then enter **4** in the third line.

> Excel will only get to this question if it has answered False to all the 4 questions, in which case A2 was >0.9, and so you want to show 4.

5. Follow steps 2–4 for the departures, using cell B2 as the basis – you must not use cell A2 for it, or the number of arrivals and departures will not be independent of one another.

6. Now you need to tell Excel to calculate the queue length at the start of the next time period. You can't have a negative queue length, so you need to use a simple **If** statement:
   If C2 + D2 – E2 > 0, you want the value C2 + D2 – E2 in cell C3. If it isn't, you want to show the value 0, like this:
   ✦ In cell C3, click on the **function** button.
   ✦ Enter **C2+D2-E2>0** in the first line, **C2+D2-E2** in the second and **0** in the third.

These steps will create a simulation looking something like this:

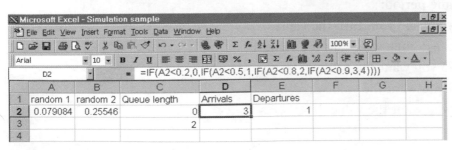

You can copy these formulae down the columns, and then draw a graph showing the behaviour of the queue:

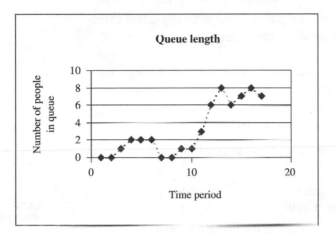

Note: The intermediate values have no meaning, so a solid line is not appropriate.

In the example, the average number arriving is 1.7 and the average being served is 1.9, so the queue will not grow uncontrollably.

However, as the graph shows, there will be periods when the queue is a substantial size for a period of time.

Supermarket managers use sophisticated models, based on these very simple ideas, to plan strategies for how many checkout counters or phone lines need to be manned at different times of the day so that customers do not encounter long waiting times.

Note: You could base a 'dice throw' random generator on the **IF** structure used in Example 2, by comparing the **RAND** value with 1/6, 2/6, 3/6 …

Exercise 9D

1. Run the simulation from Example 2. You will get different results to the example shown but you may get a similar outcome.

   ✦ Find the average arrivals and departures.

   ✦ Draw a graph to show the data.

   ✦ Interpret your graph.

2. Explore a range of Fairground games based around dice or coin tosses, for example:

   ✦ Pay 10p to toss 4 coins. Win £1 if all four show Heads.

   ✦ Will the stallholder make money doing this? Run the simulation a number of times and record how often they have to pay out.

   ✦ On average, how much does the stallholder win or lose in each game played?

- ✦ Can you give an explanation as what the average would be over a very large number of games? [Hint: you need to think about the probabilities of 'winning 90p' and 'losing 10p'.]

- ✦ If the game is changed to rolling 3 dice, and winning if 'all three show the same number', how does this compare?

3. Investigate runs of Heads and Tails instead of throws of a dice. You may find it surprising to see how often quite long runs occur.

4. Construct a simulation for the National Lottery:

- ✦ Ask the 'player' to enter their choice of 6 numbers into specified cells.

- ✦ Your simulation needs to generate 6 main numbers, plus a bonus, into another group of cells.

- ✦ This should be constructed in such a way that it is not possible for the same number to be chosen twice by the player, or the 'computer draw'.

  Don't worry about this part to start with, but once you have the rest of it working, come back and see if you can put this condition in.

- ✦ Then count how many matches there are.

- ✦ Display the prize won for that number of matches.

5. Once you have the simulation in question 4 running, look at what you could do to make the 'game' more attractive – what about the use of colour for different parts of the sheet you want to draw attention to etc.? What text should appear, and where, to make the game 'user-friendly'? – so that people could use it without needing you to explain what it is for, and what they need to do.

6. Go back to the queuing simulation in Example 2.

Note: you will need to run each simulation a number of times to know if it is acceptable – it is not enough that it seems OK the first time you run it, it needs to be repeatable.

- ✦ Change the probabilities so that the average number being served is higher. How high does the average need to be before you feel the queue behaves in an acceptable fashion?

- ✦ Is it only the average which matters, or does the balance of the probabilities make a difference? Try probabilities of 0.5 for 1 and 0.5 for 3 [average of 2 persons served] and compare it with the behaviour of the queue when the probabilities are 0.2 for 0, 0.6 for 2 and 0.2 for 4 [which also has an average of 2 persons].

Exam practice paper: Foundation tier

1. Ruby is studying snakes for a project on endangered species.
   Give an example of each of the following variables connected with
   snakes.

   (a) A continuous variable.   *(1 mark)*
   (b) A discrete variable.   *(1 mark)*
   (c) A qualitative variable.   *(1 mark)*

2. A person is chosen at random.
   The events X, Y, Z are defined as:

   X   The person was born after the year 1999.
   Y   The person was born on a weekday.
   Z   The person was born in March.

   Draw a probability scale:

   0 ——————————————— 1

   Mark on it X, Y and Z to show the probability that each event
   could represent.   *(3 marks)*

3. The bar chart represents the expenditure of the Design Technology
   department in a school.

   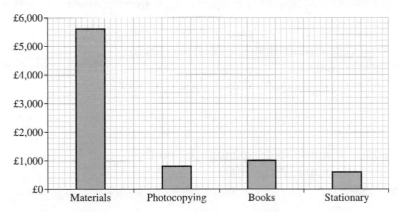

   (a) How much did Design Technology spend on materials?   *(1 mark)*
   (b) How much did the department spend altogether?   *(2 marks)*
   (c) Draw a pie chart to show the expenditure of the Design
       Technology department.   *(5 marks)*

4. A shop labels its kids clothes in one of two ways, by a letter or by a number.

Size	XXS	XS	S	M	L	XL	XXL
Height (cm)	99–102	102–112	112–124	124–140	140–145	145–152	152–157
Waist, Regular (cm)	52–55	53–56	56–60	60–62	61–64	64–66	66–69

Size	3	4	5	6	7	8	10	12	14	16
Height (cm)	91–99	99–107	107–114	114–124	124–135	135–140	140–145	145–152	152–157	157
Waist, Regular (cm)	52–53	53–55	55–57	57–58	58–60	60–61	61–64	64–66	66–69	69–71
Waist, Slim (cm)	47–48	48–50	50–52	52–53	53–55	55–56	56–58	58–61	61–64	64–66

Chloe and Ellie wanted to buy a dress.

(a) Chloe is 95 cm tall. What letter size dress should she buy?   (*1 mark*)

(b) Ellie is exactly 35 cm taller than Chloe, what letter size dress should she buy?   (*2 marks*)

(c) Ellie also wanted to buy a pair of trousers. Her waist size is 55 cm. What size trousers should she buy?   (*1 mark*)

5. Donna is carrying out a survey to find what type of fast food people prefer. She stands next to a fish and chip shop and asks customers the question:

    'Do you think fish and chips is good value for money?'

(a) Why is Donna's sample of people likely to be biased?   (*1 mark*)

Donna decides to carry out her survey amongst office workers in her town.

(b) Describe a better method of sampling that Donna could use.   (*2 marks*)

(c) Why is Donna's question biased?   (*1 mark*)

(d) Write a question to find preferred fast food as it should appear on a questionnaire.   (*2 marks*)

(e) Suggest two reasons for Donna to carry out a pilot survey.   (*2 marks*)

6. The table gives four categories of household expenditure and the price index for 1997 for three of them (1996 = 100).

Category	Index
Food	110.2
Clothing	107.5
Heat and light	
Housing	98.0

(a) The price index for heat and light increased by 7.9%. State the price index for heat and light.   (*1 mark*)

(b) Which category was more expensive in 1996 than in 1997?   (*1 mark*)

(c) If the food bill for a household in 1996 was £2200, how much would you estimate they spent in 1997? *(2 marks)*

(d) Lauren spent £344 on her summer clothes in 1997. If she bought similar items in 1996, how much would you estimate she spent? *(2 marks)*

7. The speeds of 200 vehicles travelling along a stretch of road are summarized in the table below.

Speed, v (kmph)	Number of vehicles
$30 \leqslant v < 50$	12
$50 \leqslant v < 65$	18
$65 \leqslant v < 80$	78
$80 \leqslant v < 95$	54
$95 \leqslant v < 115$	23
$115 \leqslant v < 130$	12
$130 \leqslant v < 160$	3

(a) What is the modal class speed. *(1 mark)*

(b) Draw a cumulative frequency diagram to represent these data. *(5 marks)*

(c) Use your diagram to estimate how many cars were travelling at speeds greater than (i) 48 kmph, (ii) 112 kmph. *(2 marks)*

(d) Use your answers to (b) state the likely speed limit for the road. Give a reason for your answer. (48 kmph = 30 mph, 112 kmph = 70mph) *(2 marks)*

8. A quick crossword had 25 clues. Luke drew tally charts to show the length of each answer.

Across

Answer length	Tally
5	𝍸
6	||||
7	|||
8	|
9	

Down

Answer length	Tally
5	𝍸
6	||||
7	||
8	
9	|

(a) Design a two-way table and use it to display these data. *(5 marks)*

(b) Find the range of answer length for the clues. *(2 marks)*

(c) Calculate the mean answer length for the clues. *(3 marks)*

On the same day the cryptic crossword had a mean answer length of 7.5 letters and a range answer length of 7 letters.

(d) Make two comparisons between the answer length of the clues in the quick and cryptic crosswords. *(2 marks)*

9. Suggest two ways in which the presentation of this graph can be improved.

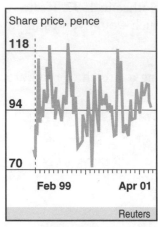

Share price, pence

118

94

70

Feb 99   Apr 01

Reuters

(2 marks)

10. (a) State one advantage and one disadvantage of telephone surveys.   *(2 marks)*

(b) A telephone directory consists of 500 pages. No page contains more than 350 names. Explain how you could use random numbers to obtain a simple random sample.   *(3 marks)*

11. Fifteen 8-year-old boys took part in a reaction-timing experiment. Their results in seconds were

$$6.0, \quad 5.2, \quad 4.4, \quad 7.7, \quad 7.2,$$
$$7.2, \quad 8.3, \quad 7.5, \quad 6.9, \quad 3.1,$$
$$8.2, \quad 5.5, \quad 7.3, \quad 6.1, \quad 7.2.$$

(a) Draw a stem and leaf diagram to represent these data.   *(4 marks)*

(b) Use your diagram find the median and upper and lower quartile times.   *(3 marks)*

(c) Draw a box and whisker diagram to represent the reaction times of the boys.   *(3 marks)*

(d) Comment on the skewness of the data.   *(1 mark)*

A group of 8-year-old girls took part in the same experiment. Their median and interquartile reaction times were 6.5 seconds and 2.8 seconds respectively.

(e) Compare and contrast the reaction times of the girls and the boys.   *(2 marks)*

12. Joss began swimming training twice a week. He recorded his time, in seconds, for 25 m breaststroke at the start of training and kept a record of his subsequent 25 m breaststroke times at time trials.

Time trial	Oct. 1999	Jan. 2000	Mar. 2000	May 2000	Sept. 2000	Nov. 2000	Jan. 2001
Time (s)	30.43	26.10	27.96	27.01	25.19	24.50	24.50

(a) Draw a scatter diagram to show this data. *(2 marks)*
(b) Comment on the correlation shown by your graph. *(1 mark)*
(c) Which time do you think has been recorded incorrectly? *(1 mark)*
(d) Draw a line of best fit on your diagram. *(1 mark)*

Joss will be awarded a bronze badge for swimming 25 m breaststroke when his time is 24.3 seconds.

(e) Use your line to estimate when Joss may gain his bronze badge. *(1 mark)*
(f) Would it be sensible to extend your trend line much further? Give a reason for your answer. *(1 mark)*

13. Mo believed he was telepathic. To test this Mo sits in a room and concentrates on a colour chosen from a choice of four, green (G), orange (O), red (R) and yellow (Y).
His sister Sadie sits in a different room and is asked to choose the colour that her brother is concentrating on.

(a) If they are guessing, what is the probability of getting a matched pair? *(1 mark)*
(b) In 200 attempts, how many matching pairs would you expect them to get if they were just guessing? *(2 marks)*

The results of the first twenty attempts are shown in the table:

Attempt	1	2	3	4	5	6	7	8	9	10	11	12	13	14	15	16	17	18	19	20
Mo	G	O	R	G	O	R	R	Y	Y	O	O	O	R	G	Y	Y	R	O	G	Y
Sadie	Y	O	G	R	O	G	O	R	Y	G	R	Y	G	R	Y	O	G	R	G	O

(c) Estimate the probability of getting a matching pair using the results of
  (i)   the first five attempts, *(1 mark)*
  (ii) the first ten attempts, *(1 mark)*
  (iii) all 20 attempts. *(1 mark)*
(d) Use your results to part (c) to comment on Mo's belief that he is telepathic. *(2 marks)*

14. Peter was training for the London to Brighton cycle ride. Peter wore a wrist heart monitor to measure his pulse during training.

(a) What method is being used to collect the data? *(1 mark)*

Peter noted the reading on the wrist heart monitor at five-minute intervals.

Time (minutes)	5	10	15	20	25	30	35	40
H (heart rate)	130	134	142	144	146	148	145	140

(b) Plot these data. *(2 marks)*

(c) In which time interval did Peter's heart rate appear to be
increasing at the greatest rate? *(1 mark)*

Peter was advised that during training his heart rate should not go
below 120 and should not go above 150.

(d) What additional information would you need to be certain that
Peter's heart rate stayed within these limits? Give a reason for
your answer. *(2 marks)*

15. The table gives the market price quarterly figures for the United
Kingdom domestic expenditure on goods and services.

Year	Quarter	Domestic expenditure (£ millions)
1990	1	82.9
	2	85.0
	3	89.6
	4	92.9
1991	1	85.8
	2	89.3
	3	94.5
	4	98.3
1992	1	*data unavailable*
	2	94.2

(a) Draw a time series graph to represent these data. *(2 marks)*

(b) Draw a trend line by eye. *(1 mark)*

(c) Use your trend line to estimate the domestic expenditure for
the first quarter of 1992. *(1 mark)*

(d) Suggest a reason for the seasonal variation apparent in the
fourth quarter of each year. *(1 mark)*

Total (*100 marks*)

Exam practice paper: Higher tier

1. A machine packs sugar into bags marked 2 kg. Samples of the bags are taken hourly and weighed. The first sample is taken at 10.00. The means of the samples are 1.99 kg, 1.995 kg, 2.005 kg, 2.105 kg, 2.2 kg, 2.305 kg.

  (a) Draw a quality control graph. *(2 marks)*

  (b) Comment on the variability shown and state what action, if any, should be taken. *(2 marks)*

2. A group of 14-year-olds take part in a reaction-timing experiment. Their results are shown below.

```
      Girls                    Boys
        8   4 | 8 | 2  3
  9  9  5  3  2 | 7 | 2  2  2  3  5  7
        5  2  0 | 6 | 0  1  9
        6  1  1 | 5 | 2  5
        3  2 | 4 | 4
             | 3 | 1
```
Key. 3 | 1 means $\frac{31}{100}$ seconds

  (a) Find the median and upper and lower quartiles for the boys. *(3 marks)*

The median and upper and lower quartiles for the girls are $\frac{65}{100}$, $\frac{79}{100}$ and $\frac{51}{100}$ seconds respectively.

  (b) On the same scale, draw box and whisker diagrams for the girls and the boys. *(5 marks)*

  (c) Comment on any similarities and differences between the reaction times of the girls and boys. *(2 marks)*

3. Ms Babs, the new headteacher of an 11–18 secondary school, decided to change the school uniform. To decide on colour and style she wanted to know the preferences of pupils at the school.

  (a) Give one reason why a pilot survey should be carried out. *(1 mark)*

  (b) Give a reason for not including sixth formers in the survey. *(1 mark)*

The numbers of pupils in years 7 to 11 are:

Year	7	8	9	10	11
Number	134	134	112	112	108

  (c) If a stratified sample of 100 is to be surveyed, calculate how many pupils in each year group you would need to ask. *(5 marks)*

  (d) Explain why it would be difficult to choose a sample of 150 stratified by age. *(3 marks)*

4. Pie charts are to be drawn to compare the funds allocated to a school department in successive years.
   In 2000, Design Technology receives £8000.
   In 2001, the department receives £8500.
   If the pie chart in 2000 has a radius of 6 cm, what should the radius be of the pie chart for 2001? *(3 marks)*

5. Thomas, my tabby cat, spends his days in the kitchen or on the bed. I am at home 4 days of the week. When I am at home Thomas is four times more likely to be on the bed than in the kitchen. If I am not at home during the day Thomas is in the kitchen.

   (a) If I am at home, what is the probability that Thomas is on the bed? *(1 mark)*
   (b) Draw a tree diagram to show the probabilities of where Thomas spends his days. *(3 marks)*
   (c) Calculate the probability that Thomas will spend his day in the kitchen. *(3 marks)*

6. A finance analyst rated the performance of ten banking institutions. (The actual name was withheld.) The table below gives these ratings and the amount of assets, to the nearest £500million, held.

Banking institution	Performance rating	Assets (£m)
A	1	14800
B	2	51000
C	3	19500
D	4	8500
E	5	12000
F	6	8500
G	7	20000
H	8	34000
I	9	20500
J	10	7000

   (a) Calculate a value for Spearman's coefficient of rank correlation for these data. *(6 marks)*
   (b) Comment on the relationship between assets and performance rating. *(1 mark)*
   (c) If the actual assets were known, what effect, if any, would this have on your calculation of rank correlation?

   Is this likely to change your answer to part (b)? Give a reason for your answer. *(3 marks)*

7. The energy, E required to make an object move at different velocities, v, is believed to be given by the equation $E = 0.5 mv^2$ where m is the mass of the object.

An experiment was carried out to test this theory. The results were:

E	2	4	6	8	10	12	15	20
v	0.9	1.3	1.6	1.8	2.0	2.2	2.5	2.9

(a) Draw a scatter diagram of E against v^2.   *(4 marks)*
(b) Draw a line of best fit.   *(1 mark)*
(c) Use your line to find the mass of the vehicle used in the experiment.   *(3 marks)*

8. A tetrahedral dice, lettered A, B, C, D, is known to be biased to the letter D.

(a) Using this extract from a table of random numbers starting at the second row,

> 76218 78176 87146 99734 74782 91613 53259 63858
> 16324 97243 03199 45435 20025 16022 27081 00058

simulate 20 tosses of the dice by letting 0 and 1 correspond to A, 2 and 3 correspond to B, 4 and 5 correspond to C, and 6, 7, 8 and 9 correspond to D.   *(3 marks)*

(b) Use your simulation to estimate the probability of obtaining the letter D.   *(1 mark)*
(c) Use simulation and the first row of the table to estimate the probability of obtaining the letter D.   *(2 marks)*
(d) Compare and comment on your answers to parts (b) and (c).   *(2 marks)*

9. The speeds of 200 vehicles travelling along a stretch of road are summarized in the table below.

Speed, v (kmph)	Number of vehicles
$30 \leqslant v < 50$	12
$50 \leqslant v < 65$	18
$65 \leqslant v < 80$	78
$80 \leqslant v < 95$	54
$95 \leqslant v < 115$	23
$115 \leqslant v < 130$	12
$130 \leqslant v < 160$	3

(a) Draw a histogram to represent these data.   *(5 marks)*
(b) Use your diagram to estimate how many cars were travelling at speeds greater than (i) 48 kmph, (ii) 112 kmph.   *(6 marks)*

(c) What is the likely speed limit for the road?
Give a reason for your answer. (48 kmph = 30 mph, 112 kmph
= 70 mph) *(2 marks)*

10. Sixty first class and forty second class letters were posted. The
number of days they took to arrive is summarized in the table.

Number of days to arrive	First class	Second class
1	48	4
2	6	22
3	3	12
4	2	0
5	1	2

(a) Calculate the mean and standard deviation of the number of
days it took the first class letters to arrive. *(5 marks)*

The mean and standard deviation for the second class letters are
2.35 and 0.853 days respectively.

(b) Compare and contrast the time it took for the first and second
class letters to arrive. *(2 marks)*

11. As part of a survey the following data on the number of children in
120 families was recorded.

Number of children	Number of families
1	16
2	35
3	25
4	18
5	14
6	9
7	3

(a) Write down the median number of children per family. *(1 mark)*
(b) Draw a step cumulative frequency polygon to represent these
data. *(4 marks)*
(c) Find the interquartile range of the number of children per
family. *(2 marks)*
(d) Explain why the interquartile range is only an approximate
measure of spread for this distribution. *(1 mark)*

12. The table gives four categories of household expenditure, the weight assigned to each category and the price index for three of them.

Category	Index	Weight
Food	104.2	150
Clothing	103.5	60
Heat and light		45
Housing	98.0	170

(a) The price index for heat and light increased by 7.9%. State the price index for heat and light. *(1 mark)*

(b) Calculate a weighted price index for these items of household expenditure. *(4 marks)*

(c) What does this index tell you about overall household expenditure? *(1 mark)*

13. Mo conducted an experiment to see if he was telepathic. He telepathically tried to communicate one of four different colours.

(a) Write down the probability of correctly identifying the colour. *(1 mark)*

Mo did this five times.

(b) Give two reasons why the binomial distribution is a suitable model for the experiment. *(2 marks)*

(c) Copy and complete the table of probabilities: *(5 marks)*

Number correctly identified	0	1	2	3	4	5
Probability	0.237	0.396	0.264			

(d) If Mo carries out the whole experiment on twenty different people, how many of those people could be expected to identify 2 colours correctly? *(2 marks)*

14. The table gives the market price quarterly figures for the United Kingdom domestic expenditure on goods and services.

Year	Quarter	Domestic expenditure (£ millions)
1990	1	82.9
	2	85.0
	3	89.6
	4	92.9
1991	1	85.8
	2	89.3
	3	94.5
	4	98.3
1992	1	90.6
	2	94.2

(a) Draw a time series graph to represent these data. *(2 marks)*
(b) Calculate appropriate moving averages. *(3 marks)*
(c) Plot these moving averages and draw a trend line. *(3 marks)*
(d) Use your line to predict the domestic expenditure for the third quarter in 1992. *(3 marks)*
(e) (i) Calculate the average seasonal variation for the first quarters of the years. *(3 marks)*
    (ii) Hence estimate the domestic expenditure for the first quarter of 1993. *(2 marks)*

Total (*120 marks*)

Answers

1 Data collection

Check in 1

1. (a) i) 23 ii) 4, iii) 10; (b) i) 卌| ii) 卌 卌 |||| iii) 卌 卌 卌 卌 iv) || v) 卌|
2. (a) 4 (b) 19 (c) 23 (d) 11 (e) 33 (f) 67
3. (a) $\frac{6}{40}$ (b) $\frac{10}{40}$ (c) $\frac{3}{40}$ (d) $\frac{8}{40}$ (e) $\frac{11}{40}$ (f) $\frac{16}{40}$ (g) $\frac{33}{40}$

Exercise 1A

1. (a) Discrete (b) colour.
2. Texture of hair.
3. Continuous.
4. (a) (i) colour (ii) length
   (b) (i) type of fish (ii) number in the shoal
   (c) (i) how realistic (ii) playing time.
5. (a) (i) position (ii) time taken
   (b) (i) postage cost (ii) parcel weight
   (c) (i) eggs (ii) sugar.
6. Descriptive, judgemental, assigned a number rather than countable.
7. Discrete.

Exercise 1B

1. Secondary.
2. (a) Secondary (b) Jason, more data used.
3. Primary, need to know if people actually like/will buy it.
4. Use hand span data so that kettle is comfortable for majority of population to hold.
5. Explanatory – leg length, Response – time taken.
6. Personal interview, needs to find out what people want/would use.
7. Only need to ask people living in new houses.

Exercise 1C

1. All of this type of elastic band. Test to destruction.
2. Need to be aware of what workers etc would want and population is small, population is everyone who works there.
3. (a) Sample, too many to look at every house. (b) Census, small number, likely to be individual in a village.
4. Everyone living within range of floodlight beam plus everyone that may be affected by increased traffic/parking at evening matches.

Exercise 1D

1. (a) Town residents in work, (b) likely to travel by rail, (c) random.
2. (a) Children at the school, (b) not in proportion, (c) stratified.
3. (a) Whole population, (b) one area, possibly unemployed, (c) stratified (or opinion poll).
4. (a) Population of Britain, (b) one area, (c) stratified.
5. (a) Married people in Scotland, (b) Churchgoers more likely to marry in church, (c) systematic.
6. (a) All schoolchildren, (b) only one school, (c) stratified by age/gender (or clusters of other schools).
7. (a) All bulbs manufactured by Glowalot, (b) tests to destruction, (c) systematic.
8. (a) All smokers, (b) poor people may not be able to afford return postage/does not include the homeless, (c) cluster.
9. (a) The area of lawn, (b) only one small part of the lawn, (c) cluster.
10. (a) Split pins produced by the machine; (b) if every ten may all be defective or none defective; (c) random.

Exercise 1E

1. 'A lot' is undefined. Give choices of number of hours per day.
2. Leading. Rewrite with choices poor to very good.
3. Mix of units plus gap i metre and between five feet and 2 metres. Re do choices with no gaps.
4. (b) should appear before (a) and (a) should read 'If yes , what do … ' and have choices.
5. Young, middle-aged, old are undefined. Replace with age ranges.
6. Gaps in choices. Replace with 0, 1–3, 4–6, daily.
7. Questionnaire should include age/year group and preferred flavours of crisps. Whole school is population. Sample should be stratified and about 10% of total. Pilot survey could identify which flavours to include.
8. Pilot questionnaire should include open questions to see which features people identify and how much they would pay and if they buy one. (a) Whole population could be sampling frame. (b) Random or stratified.

Exercise 1F

1. Matched pairs.
2. Data logging.
3. Data logging.
4. Control group.
5. Capture – recapture method.
6. Before-and-after.
7. Capture – recapture method.
8. 300.
9. 400.

Exercise 1G

1. (i) Class tally chart for pets, (ii) Class tally chart for bicycle colour.
2. Frequencies: 2, 5, 2, 2, 3, 3, 2, 1
3. Tally chart with 6 rows (for the numbers on a dice).
4. Tally chart with 5 rows (for each vowel).

Exercise 1H

1. £120.50; £119.50.
2. £125; £115.
3. 750.5g; 749.5g.
4. 755g; 745g.
5. 1010g (1.01kg); 990g (0.99kg).
6. 47.5km; 42.5km.
7. 26.5cm, 25.5cm if nearest cm; 26.05cm, 25.95cm if nearest tenth cm; ...
8. 40.5m, 39.5m (nearest m); 42.5m, 37.5m (nearest 5m); . . .
9. 9.75m, 9.65m (nearest 10cm); . . .
10. 35.5g, 34.5g (nearest g); . . .

Exercise 1I

1. Frequencies: 6, 8, 10, 9, 7
2. Frequencies: 4, 4, 11, 5, 4
3. Frequency table using class data.

Exercise 1J

1. (a) 37   (b) 7   (c) 8.
2. Almost double the next heaviest coin.
3. Two-way table, number of brothers & number of sisters; own data.
4. Two-way table, gender & types of music; own data.
5. Two-way table, colour & weight; own data.

Check in 1

1. (i) colour or design, (ii) number of ring tones or size of phone.
2. Stratified sample asking 2 girls for every 3 boys asked.

3. Put the following features in order of preference, 1 = most wanted feature (provide a list of features).
4. Ask pupils to sit in a quiet place with their eyes closed. Ask them to say Now 30 seconds after you say Start Timing.
5. Tally chart with appropriate time intervals.

Revision Exercise 1

1. (i) Qualitative; (ii) Continuous; (iii) Discrete; (iv) Qualitative.
2. Discrete; Continuous; Discrete.
3. (a) Primary; (b) Number of children, amount spent; (c) Journey time.
4. (i) Weight of an ingredient or cooking time; (ii) number of eggs or lemons; (iii) Taste.
5.

	Classical	Pop
Under 25		
25 & Over		

6. (a) Any table with 7 appropriately labelled compartments; (b) Any table with 6 compartments for the 6 customers.
7. (a) Frequencies 3, 10, 5, 2; (b) 4.
8. Small sample; all from one family (biased).
9. (a) Check wording, find typical views; (b) Number each person, put numbers in a hat and select 130; (c) $\frac{780}{3900} = 0.2$; sample 13, 40, 259, 468.
10. Method 2; sampling frame involves (almost) all groups of people (adults).
11. (a) Quicker/cheaper; (b) All likely to want it, limited sampling frame; (c) Random sample sampling frame everyone within local area, (d) Suitable question with at least 3 non-overlapping choices with no gaps.
12. (a) Census includes whole population, sample only some; (b) Test to destruction; (c) Census too large/ saves time/ saves money.
13. (a) Questions are short/relevant/closed/not leading, personal/unbiased/easy to understand, answer;
    (b) (i) Large coverage/fewer staff/cheaper/answer questions at leisure; (ii) non return/loss/no help given;
    (c) (i) Personal/do not like giving weight; (ii) Clear wording, 3 non-overlapping groups, no gaps.
14. A Personal/ambiguous, rewrite with choice boxes to tick;
    B Leading, rewrite 'What do you think ...' & include tick boxes or opinion scale to mark.
15. Where did you go on holiday last year? Britain [ ] Europe [ ] Elsewhere [ ] Did not go on holiday [ ].
    If you went on holiday last year, was your accommodation: Hotel [ ] Self-catering [ ] Camping [ ] Other [].
16. Measure a sample of people in the morning and again that same evening, using same means of measuring.
17. (a) Not random allocation, only given to girls, no control, only children, only his patients; (b) Use a control group, random allocation, equal numbers boys and girls.
18. (a) May be other reasons that people lose/do not lose weight; (b) Paired same gender, similar weight, height, lifestyle etc; (c) To keep all other factors the same.
19. (a) 10; (b) All 20; (c) 10; (d) 6/10; (e) Ensure truthful answers.
20. (a) $\frac{20}{50}$; (b) High non-response, short people less likely to respond;
    (c) Tick the box giving your height. At least 3 non-overlapping and no gaps alternatives.
21. (a) Names in hat, choose 80 or number all 500 students and select 80 using random numbers; (b) 80/500 = 0.16, sample 16, 12, 20, 16, 16; (c) Interviewer chooses (appropriately classified members from each year) up to the required number.
22. (a) Time short, Sample small, Business people unlikely at 11am, No criteria for choice, Interviewer bias;
    (b) (i) Number and select every nth; (ii) Divide population by some criteria, select in proportion to total;
    (iii) Stratified ensures selection across each group; (c) In which age group are you? (3 groups not overlapping, no gaps).
23. (a) Quick/cheaper/easier; (b) All men/women/same floor; (c) 20, 10, 20; (d) 4; (e) Number 1 to 20, choose every 5th random start 1–5 .
24. (a) 1 – 30 in hat pick 5 or number 1 30 use random number tables ignore repeats; (b) Every 6th number random start 1–5; (c) Systematic, not clustered in one part of the street.
25. (a) P Taylor, D Peters, B Quarishi; (b) All boys; (c) Stratified by gender.

2 Interpretation of data

Check in 2

1. $\frac{3}{10}, \frac{21}{50}, \frac{1}{2}, \frac{3}{4}, \frac{89}{100}$
2. *Drawings of scaled lines*
3. (a) 17:13, (b) 3:2, (c) 9:4, (d) 3:2

Exercise 2A

1. (a) £1, (b) 5p, 10p, 20p, 50p; British silver coins
2. (a) 35%, (b) Ireland, (c) 11%
3. (a) Shredded Wheat, (b) Rice Krispies, (c) Shredded Wheat or Shreddies
4. (a) gymnastics, swimming (hard) or Yoga, (b) rowing & swimming (hard), (c) tennis & badminton,
   (d) Very good (instead of beneficial) strength

Exercise 2B

1. People who garden as a hobby are more likely to grow vegetables.
2. Half, 50%, is average. 37% less than half.
3. (a) You do not know where those questioned live. May be hoping for increased business.
   (b) Air traffic also increased, need to know the proportion of fatalities.
   (c) Unaware of how many expressed a preference or who took part in the survey.

Exercise 2C

1. (a) thick line, drop shadow, false zero; (b) scale does not begin at zero; (c) years not linear; (d) no scale given.
2. No, need scale on vertical axis.
3. Do not know what is represented on horizontal axis, data may not be recorded at equal intervals.
4. Zenith Suppliers, sales gradient £4200 (Arcwell £75).
5. (a) Vertical axis begins at 100000 and is not scaled; (b) Impression given of greater increase; (c) Actual sales or vertical scale.
6 First graph gives an impression of greater variation in sales and Spring sales appear almost negligible.

Exercise 2D

1. (a) (i) 1:3, (ii) 1:9, (b) No, area of picture more prominent than the values.
2. Have their areas in proportion to the total sales.
3. Horizontal scale not linear. Height of wheelie bins is difficult to read. The 3-D picture gives a different impression to the height. Only reported non-collections, what about non-reported?

Check out 2

1. (a) Exotic fruit, (b) Garden berry.
2. Vertical scale does not begin at zero; horizontal scale is not linear.

Revision Exercise 2

1. (a) Income Tax; (b) £21.7 billion; (c) £23.1 billion.
2. 50% is average therefore 40 % less than half is less than average.
3. (a) 60%; (b) 1995; (c) 20%; (d) North; (e) 88p; (f) 68p.
4. No horizontal scale/labels; 92.62 higher than 96; Width of line so large that exact coordinates are obscure; Thick line (17.48 looks more like a 10); Shadow gives two positions of peaks.
5. (a) 6%; (b) 32–34; (c) Any difficulty less than individual column entries; (d) $3 + 2 + 3 < 14$.
6. Misleading title; false zero; sloping lines; unequal distances between numbers on horizontal scale.
7. False zero; suggests that growth has tripled when not actually doubled; do not know about intervening years.
8. (a) $109230 m; (b) $88073 m (a deficit); (c) $142613 m (a deficit).
9. Areas not proportional to data.
10. (a) Ratio of areas is 1 : 4; (b) 5.04 cm.

3 Tabulation and representation

Check in 3

1. (a) 2.8
4. (a) 20   (b) 1.2   (c) 20   (d) 0.4
5. (a) 15   (b) 45   (c) 62.5%

Exercise 3A

1. (b) Fourton   **2.** 400, 300, 150, 650   **3.** (b) Statistics and mathematics

Exercise 3B

1. (a) Walk   (b) 26

Exercise 3C

2. (a) 1   (b) 3   (c) 4

Exercise 3D

1. (a) 40   (b) (i) 39   (ii) 13

Exercise 3E

1. (b) 14.1 cm   **2.** 4.62 cm   **3.** 8.95 cm   **4.** 42.75 million

Exercise 3F

1. (a) 6, 19, 36, 43, 47, 49, 50
2. (a) 9, 21, 39, 60, 77, 92, 100
3. (n) 2, 8, 22, 40, 50, 56, 60

Exercise 3G

1. Ecuador, French Guiana

Exercise 3H

2. (a) bar chart   (b) colour is quantitative
3. (a) bar chart or vertical line graph
   (b) data is numerical and there are too many items for a pie chart

Exercise 3I

2. (b) an increase   (c) rectangular

Exercise 3N

1. Symmetrical   **2.** Almost symmetrical   **3.** Positive skew
4. Slight positive skew   **5.** None of the descriptions fit

Check out 3

2. (a) bar chart   (b) pie chart

Revision Exercise 3

1. (a) 15%   (b) 39%   (c) 65%   (d) Austria
2. (a) 27   (b) £220 000
3. (a) 90   (b) 34
4. (a) (i) 16.5%   (b) (ii) 14.7%
5. (a) 2700   (b) 900, 1300, 1500, 1200, 700, 300, 100   (c) 26 years   (d) 19 years
7. (a) 86   (b) $213\frac{1}{3}$ (c) 11.5 per year
8. (a) 80, 60, 40, 60
9. (b) (i) $\frac{1}{2}$ year (ii) 23.5   (c) (i) 48 (ii) 174°   (d) 2:3
10. 90°, 80°, 100°, 30°
11. (a) 160 million   (b) 400 million
12. (b) $14 \leqslant x < 15$ (c) 0.1002
14. (a) 65°, £90 million
15. (a) 6   (b) 26   (c) 34
16. (a) £60–£75   (b) £45–£60
17. (a) Positive   (b) 20   (c) 41 minutes

4 Measures of location and spread

Check in 4

1. (a) 0.35   (b) 0.43   (c) 0.107
2. (a) 49.5, 59.5   (b) 50, 60
3. 7

Exercise 4A

1. (a) 3, 4, 4.89   (b) 15, 15, 15.2   (c) 23, 23, 22.57
2. 46, 52, 51.7; 39, 52, 56.2
3. (b) 3, 4, 5, 5, 5.50, 7.50, 8, 8, 10; £5.50   (c) £6.22   (d) £6.40, £6.50, £8
4. (a) 1320 cm   (b) 165.33 cm

Exercise 4B

1. (i) No mode, 7.5, 7.75   (ii) No mode, 5.8, 5.375
2. 67
3. 1.72
4. 2.8 kg
5. 6, 7, 7.02
6. 35, 35, 34.525
7. 4, 4, 4.57

Exercise 4C

1. 27.87
2. 5.25
3. 40
4. (a) 515.3

Exercise 4D

1. (a) 9   (b) 34.17   (c) 14.62
2. (a) 120 to 130   (b) 123.2 m   (c) 6, 19, 36, 48, 50   (d) 123.5 m   (e) 150.5 m, 100.5 m, 50 m, 30 m

Exercise 4E

1. (a) 1003.6   (b) 510.25   (c) 82.3
2. 1009.90
3. 443.75

Exercise 4F

1. (a) 15
2. 26.55
3. Denim
4. (a) 30.57
5. Median 1016
6. Mode
7. Mean 15.15
8. median 21

Exercise 4G

1. (a) (i)   6   (ii) 4   (iii) 7.32
      (b) (i)   7.5   (ii) 4.67   (iii) 8.25
2. (a) 5.995%   (b) 5.325%
3. 20.6%
4. (a) 13.61, 30; 25.99, 26   (b) 33.3; 26   (c) 0, 22.5; 0, 19.5

Exercise 4I

2. (a) (i)   6   (ii) 34
(b) (i)   45   (ii) 37

Exercise 4J

3. 11.8, 12.4, 13, 15, 16, 16.8, 18.8 Positive

Exercise 4K

1. (b) 23.33; 27.24; 32.67
2. (b) 152.68 − 145.33 = 7.35
3. (b) 15.36; 33.14; 46.54   (d)   Positive
4. (b) 123.9; 143.5 − 92.3 = 51.2
(c) Boris

Exercise 4L

1. about 24; about 34
2. about 142; about 157
3. 48

Exercise 4M

1. (a) 6, 2.28   (b) 16, 8.38   (c) 9, 4.80
3. (a) (i) 5, 2.66   (ii) 50, 26.6   (iii) 3, 2.66

Exercise 4N

1. 5.3; 3.95
2. 5; 2.42
3. 8.17; 4.94
4. 10.5; 2.72
5. 30.63; 8.54
6. 7.02; 2.06
7. 34.525; 1.688
8. 21.65; 6.39

Exercise 4O

1. 0.5; −1.5
2. (a) 2, −1, 1.4; 1.75; −1; 1.375   (b) Mark

Check out 4

1. mean = 5.57
median = 6
mode = 7
2. (a) range = 5
interquartile range = 2
standard deviation = 1.68
(b) mean = 59.4
standard deviation = 20.8
4. median

Revision Exercise 4

1. (a) 7, 5, 4, 2, 1, 1   (b) mode   (c) median 120, mean 120
2. (a) 2   (b) 9, 1   (c) 8
3. (b) (i)   54   (ii) 22
4. (a) 676
5. (a) 30, 35, 17, 12, 6   (b) (i) 5.3; 6.6   (c) (i) 1.06, 0.897
6. No answers

7. (a) 86   (b) 4.5   (c) 5.3, 3.7   (d) 4.2   (e) 0.0879
8. (b) 93.9s, 46.1s
9. (b) (i)   55   (ii) 43, 67
10. (b) 10.02 cm
11. (a) (i)   5.25   (ii) 5.75, 4.8
12. (a) 2.12   (b) 1.09
13. (a) 12, 39, 73, 97, 112, 120   (c) 22.25   (d) (i) median
14. (a) 34   (b) 10.9
15. (c) 18.5s
16. (a) 44   (b) 3%
17. (a) (i)   51   (ii) 2.07
18. (b) £83 444, £20 421   (c) 40   (d) (i) 1991   (ii) 165
19. (a) (i)   34   (ii) 80   (b)   308 lb
20. (b) (i)   54   (ii) 22
21. (b) £31 350, £12 121   (c) 0.558, 0.166
22. (c) (i)   43.4   (ii) 52.1   (iii) 34.1   (d) (i) 8.7, 9.3   (e) 60 good, 37 average, 28 poor
23. (a) (I)   1.5   (ii) 0.98   (b)   0.36 0.65
24. (a) (i)   £15 000   (ii) £12 000   (iii) £22 000   (iv) £10 000   (v) £20 000
25. (b) (i)   2   (ii) 6
26. (a) 1.5   (b) 1.53m
27.   No answers
28. (a) 5.5   (b) 0.764   (c) (i) 0.39   (ii) 5.7
29. (b) (i)   35   (ii) 20.1   (c)   40.25, 16.89
30. (a) 77

5 Other summary statistics

Check in 5

1. (i)   19.4 (ii)   22.
2. Square
3. (a) 8:15   (b) 30:13   (c) 19:12.

Exercise 5A

1. Bethany 66.45% (Andrew 63.85%)
2. 73.2%, no distinction
3. 63%
4. Annabel (6.35) got the post (Barbie 5.15; Cindy 5.75)
5. 6.5

Exercise 5B

1. 108.3%
2. 106.47%
3. 91.67%
4. £4.90
5. £75
6. (a) 1992; (b) Prices always increase; (c) £20.16; (d) £140; (e) 113.3
7. (a) washing machine £140, dishwasher £130; (b) WM; (c) DW; (d) Actual increase for DW greater although % increase smaller as original price greater to begin with.

Exercise 5C

1. (a) increase; (b) small decrease then increase; (c) no change; (d) increase, no change then decrease; (e) increase, decrease, no change.
2. 166.67, 125, 116, 120.69
3. 60, 69.4, 72, 80.56; annual decrease, but % decrease smaller each year; 1996 index – 30, 70% decrease in 3 years, only 28% over the previous year.
4. 100, 105, 106.67, 107.14, 113.3
5. 100, 120, 148.8, 184.5, 202.96; more than doubled.

Exercise 5D

1. 154.43
2. 82.13
3. 106.9
4. 79.54
5. 124.65
6. Gilding metal 81.43; Dutch metal 82.74; Dutch metal costs changed the most.
7. 80.61; 170.65; 91.46; 79.45; 81.69. Cost decreased, almost doubled, then decreased for the next three years.
8. 82.38; 157.36; 90.89; 85.62.

Exercise 5E

RPI for own collected data.

Exercise 5F

1. 217.5
2. (a) 150   (b) 30.
3. 496
4. 106
5. 16.8; distortion of true figure as not all the women will be of child bearing age.

Exercise 5G

1. 479.6; 23.98 pupils absent / day (assume school month = 20 days).
2. Sun City 103.57, Cloudy Town 32.34; Cloudy Town, less likely to develop skin cancer.
3. Acmebuilt 52.27; Bodgitt 88.53; Acmebuilt has a lower accident rate.

Exercise 5H

1. Time series graph & trend line; missing bill £64.
2. Time series graph & trend line; includes Christmas shopping period.
3. Time series graph & trend line; long weekend away by staff.
4. Time series graph & trend line; Trend line suggests decrease in attendance; week 3 could have coincided with school half term holiday; consider latest film releases, time of year, school holidays.

Exercise 5I

1. (a) 4-point moving averages: £47.25, £48.50, £49.63, £51.75, £53.25, £54.13, £55.75; (c) £60; (d) similar possibly more accurate; (e) £24 £25; (f) £86–£87.
2. (a) Profit is given half yearly that is two values per year; (b) Moving averages: £25500, £25500, £27500, £29250, £29750, £30500, £31750, £32500, £33250; (d) Jan. – June £29500; July – Dec. £41000.
3. (a) 5-point moving averages: 5.2, 5, 5, 5.2, 5.4, 5.6, 5.6, 5.8, 5.6, 5.4, 5.2; (d) Monday ~12; Tuesday ~4.
4. (b) 3-point moving averages: £361.67, £365, £370, £380, £381.67, £386.67, £395, £401.67. (d) ~£410; (e) ~ – £51.25; (f) ~ £373.75.

Exercise 5J

1. (a) No action; (b) General increase, machine or operator may need adjusting; (c) Last two values outliers, check machinery; (d) One value too low, check machinery; (e) Too perfect.
2. Monday extreme values at 10.00 & 11.30; Tuesday no action; Wednesday too perfect all morning; Thursday steady decrease throughout morning; Friday outlier from 11.30.

Check Out 5

1. 70.5
2. 125
3. 7
4. Time series graph trend showing increase year on year.
5. 2-point moving averages; plot 42.5, 45, 48, 50, 54.5, 57.5, 60.
6. Too low, but increasing, check machinery and/or operator. Chapter 5 Revision Exercise (Past Paper Questions)

Revision Exercise 5

1. (a) 55   (b) 90.
2. (a) $2 \times 2.5 + 4 \times 1 + 12 \times 4 + 1 \times 5 = 62$; (b) £72; (c) 116.1; (d) 33.3%; (e) Labour costs rising at a significant higher rate.
3. (a) 133.3   (b) £14.
4. (a) Cost is 5% higher than in 1994; (b) (i) 32.40; (ii) £2.40; (c) 116.
5. (a) 112.5   (b) £600   (c) 25%   (d) 11.1%.
6. (a) 125; (b) 61.8p; (c) Price increases by a factor 1.2.
7. (a) $\frac{1600}{80} \times 100 = 2000$; (b) 2100, 600, 1000.
8. (a) (i) 75; (ii) 150; (b) Same as 1988.
9. (a) Takes account of proportions; (b) £500; (c) 117.25.
10. (a) (i) $((250 \times 208) + (305 \times 77) + (358 \times 38) + (308 \times 149) + (391 \times 67))/539$; (ii) 320.7; (iii) 107.2; (b) Reduction in index value, high relative price increase, significant weighting.
11. (a) 1973, (b) Higher (expected 68); (c) Exercise / nutrition / medicine / ...
12. (a) 24   (b) 19   (c) 7505.
13. (a) 63   (b) 48 (47.95).
14. (a) 15; (b) Retirement town / More elderly people / ...; (c) 19.96; (d) Population proportion different.
15. (a) 13.4; (b) Town Y more older people; (c) 15.8; (d) Compares different towns with differing age distributions.
16. (a) 4, 2, 15, 50; (b) Westhope 13.6, Martrent 19.1; (c) Yes, lower standardized death rate.
17. (a) Plot year (x) car production (y) points (1984, 900) (1986, 1000) (1988, 1250) (1990, 1290) (1992, 1290) (1994, 1500) (1996, 1690); (b) Suitable line crossing axis between 740 and 840; (c) (i) ~930000; (ii) ~1820000.
18. (c) 23.7; (d) High, increases smaller in later years.
19. (a) 1992; (b) School closed for summer holiday; (e) Turnover is falling.
20. (b) 105, 108.25, 109.75, 112.25, 116, 116.5, 120, 123.75; (d) £129.
21. (b) Average must contain one from each period; (c) 77.3, 74.7, 100, 93.3, 97.3, 70.7, 80, 74.7, 81.3, 96; (e) Trend line or regression line.
22. (b) Weather / Longer days / Growing season / Spring enthusiasm; (c) 22, 23, 24, 24.75, 26, 26.5, 27.5, 28.5, 29, 30, 30.5, 31.5, 32.5; (e) £330 000–£370 000.
23. (b) 217, 223, 228, 235, 240, 245, 250, 256; (c) Weekly takings; (e) ~ £13500; (f) £256000–£263000.
24. (b) (i) Peak demands in Q1 and Q4; (ii) Winter seasons; (c) 1135, 1180, 1215, 1265, 1310; (f) Q1 ~ 1810, Q2 ~ 1382.5.

6 Correlation and regression

Check in 6

1. Isosceles triangle.
2. 22.08
3. (a) $y = 5x + 4$ (b) $y = -3x + 1$ (c) $y = 3x - 8$ (d) $y = -5x - 4$

Exercise 6A

1. B
2. A
3. C
4. Sketch, negative correlation, temperature (x-axis) gas used (y-axis).
5. Sketch, positive correlation, age (x) weight (y).
6. Sketch, no correlation, height (x) amount spent (y).

Exercise 6B

1. Correlation.
2. Causality.
3. Third variable.
4. Correlation.
5. Causality.

Exercise 6C

1. Scatter graph, height (x) arm span (y), mean point (144, 138.5) line of best fit with positive gradient.
2. Scatter graph, age (x) weight (y), mean point (44.1, 9.2) line of best fit with negative gradient.
3. Scatter graph, published price (x) book club price (y), mean point (10, 7..4) line of best fit with positive gradient.
4. Scatter graph, Number of absentees / number of thefts, no correlation mean & line of best fit inappropriate.

Exercise 6D

1. Strong positive correlation.
2. Strong negative correlation.
3. Strong positive correlation.
4. No correlation.

Exercise 6E

1. Answer in range 138 cm–141 cm.
2. (a) Approximately 9g; (b) Answer in range 4.8g–5.8g; (a) is more reliable, within the plotted points.
3. (a) Answer in range £21–£22; (b) approximately £8; (b) is more reliable, within the plotted points.
4. (a) £17; (b) £2; (c) £0, line of best fit has reached x-axis.
5. (i) ~3600; (ii) ~5250; (iii) 0 100; (ii) unrealistic, 12 year olds should not be driving.
6. Pregnancy unlikely to be this long, no longer premature, but very overdue.

Exercise 6F

1. $y = 4x + 3$
2. $y + 2x = 7$
3. $2y + 3x = 6.4$
4. $y = 2.9x - 3.75$
5. (a) (i) their line using one point and (144, 138.5); (b) (i) increase in arm span with height (ii) unrealistic.
   (a) (ii) their line using one point and (44.1, 9.2); (b) (i) decrease in weight with age (ii) unrealistic.
   (a) (iii) their line using one point and (10, 7.4); (b) (i) book club price factor of published price (ii) realistic if 0.

Exercise 6G

1. Plot $\frac{1}{S}$ (x-axis) T (y-axis) points (1, 89) (0.5, 46) (0.33, 30) (0.23, 25) (0.2, 20) (0.17, 16) (0.14, 13) (0.125, 11) (0.11, 10) (0.1, 9). (i) ~62 °C; (ii) ~20 °C; (iii) ~1 °C.
2. (a) Plot $\sqrt{A}$ (x-axis) C (y-axis) points (1.41, 39.3) (2, 42) (2.45, 44) (2.83, 45.4) (3, 46) (3.16, 46.4).
   (b) ~ 15cm. Unreliable outside plotted points; growth rate may change.
3. Plot S^2 (x-axis) C (y-axis) points (784, 20) (900, 21) (1225, 24) (1600, 30) (2500, 35) (3025, 42). a = 0.01; b = 12.

Exercise 6H

1. 0.29; closest to zero.
2. A 0.78; B $(-)$0.81; C 0.39; 1.32 not a coefficient as greater than 1.
3. $r_s = 1$; same rank order for each test.
4. $r_s = 0.25$; very little agreement.
5. $r_s = 0.814$; high degree of agreement.
6. $r_s = -0.579$; weak negative correlation, the longer the time the less errors; employ C quickest with not too many errors.

Check out 6

1. Plot points (40, 52) (64, 75) (36, 47) (50, 72) (72, 83); same order, but higher % in test Y.
2. Draw line of best fit through mean (52.4, 63.8).
3. $y = x + 12$
4. $r_s = -0.371$; opposing tastes in jam, if preferred by one judge then not liked by the other.

Revision Exercise 6

1. (a) As thickness increases the amount of heat escaping decreases; (b) As temperature increases in NZ the amount of coal sold in the UK increases; (c) Part (a).

2. (a) Plot (20,10) (25, 21) (30, 18) (35, 24) (40, 28) (45, 30) (50, 34) (55, 40) (60, 45) (65, 50); (b) 42.5;
(c) (i) (42.5, 30); (e) Aged 57, within the plotted points; (f) B.

3. (a) Plot (48, 53) (20, 26) (13, 17) (18, 20) (40, 38) (29, 40) (34, 37) (43, 49); (b) (i) 30.6; (ii) 35;
(c) Line passing through mean; (d) (i) 29; (ii) 64; (e) First one , interpolation.

4. (a) Plot (63, 141) (13, 14) (34, 43) (80, 170) (51, 95) (14, 21) (45, 72) (74, 152) (24, 31) (82, 171); (b) (48, 91);
(c) Line through mean and (20, 24.5); (d) (i) £179; (ii) –£11; (iii) £119.50; (e) Order (iii), (i), (ii); interpolation, extrapolation, not possible.

5. (a) 108kg; (b) 168cm; (c) Player (172, 98); (d) Positive; (e) (i) greater than 84–86; (ii) > 190 cm.

6. (a) Plot (167, 101); (b) Line through mean point; (c) ~ £62; (d) (i) Beetall, £55 gets a camera with a higher usual price; (ii) Difference is greater.

7. Plot (152,210) (165, 235) (144, 208) (169, 252) (154, 222) (151,224) (160, 230) (157, 241); (b) 227.75;
(c) line passing through mean; (d) approx 238 cm (e) H.

8. B, Perfect positive; A, Negative; C, No correlation.

9. (a) $y = 1.55x + 0.05$; (b) If straight line distance is zero then so is journey distance;
(c) For each km away from the school you have to journey 1.55km.

10. (a) Plot (7.8, 29) (8.1, 28) (6.4, 26) (5.2, 20) (7, 24) (9.9, 35) (8.4, 30) (6, 22) (7.2, 25) (10, 36); (b) 27.5;
(d) $y = 3.2x + 3$; (e) ~30; (f) Extrapolation.

11. (a) Straight line not appropriate; (b) Extrapolation not recommended, Growth has stopped.

12. (a) Plot (18, 37) (36, 54) (45, 63) (22, 42) (69, 84) (72, 91) (13, 33) (33, 49) (60, 79) (79, 98) (10, 32) (53, 70);
(b) (42.5, 61); (c) Line must pass through mean and given point; (d) (i) £83; (ii) £102; (e) (i) interpolation;
(f) 0.981; (g) per unit increase in production cost.

13. (a) Plot (1500, 9800) (1300, 8200) (1100, 7000) (1600, 11200) (1250, 7800) (1800, 12000);
(b) Line through mean (1425, 9370); (c) $y = mx + c$ where m = 7 to 8.5, c = ⁻2500 to ⁻500;
(d) (i) £11000 – £12000; (ii) Specifications vary with capacity.

14. (a) Y; (b) X, ~ ⁻0.95; Z, ~ +0.7

15. (a) ⁻0.52; (b) I, ⁻0.52 closer to ⁻1 than 0.45 is to ⁺1.

16. (a) 0.6; (b) Weak or moderate agreement.

17. (a) First judge awards higher points than the second judge; (b) (i) 0.95; (ii) Strong agreement.

18. (a) 0.714; (b) 35; (c) (i) 49; (ii) 52; (d) 1.

19. (a) Rank 1   1   2.5   5   4   2.5   6; (b) (i) C; (ii) Ranks must remain unchanged.
Rank 2   1   3.5   5   3.5   2   6

20. (a) Sales   1   2   3   4   5   6   7   8   9   10 ; (b) 0.71;
Grants   1   4   6   2   7   5   8   3   9   10
(c) Positive correlation; (d) Anything < £107.5 million.

21. (a) 0.1875; (b) Ranks for attendance figures for 1986 more similar to 1991, than 1981 is to 1991.

22. (a) 0.396; (b) Little agreement between the two jumps; (c) (i) No correlation; (ii) Strong positive correlation;
(iii) Mistake made in the calculation.

23. (a) Sunshine   4   6   8.5   10   8.5   5   3   2   7   1
Temperature   6   10   9   8   2   4.5   7   3   1   4.5
(b) 0.2; (c) Almost no correlation; (d) (i) Sunshine & temperature not related;
(ii) inverse relationship between sunshine & temperature.

7 Probability

Check in 7

1.

Fraction	Decimal	Percentage (%)
$\frac{2}{5}$	0.4	40
7/20	0.35	35
9/50	0.18	18

2. (a) (i) $\frac{19}{12}$ (b) (i) 0.12
(ii) $\frac{2}{3}$ (ii) 6.25
(iii) $\frac{5}{4}$

Exercise 7A

1. (a) 0.4 (b) 0.5 (c) 0.6 (d) 0.4
2. (a) $\frac{1}{13}$ (b) $\frac{1}{26}$ (c) $\frac{25}{26}$ (d) $\frac{3}{13}$ (e) $\frac{11}{13}$
3. $\frac{1}{25}$
4. (a) $\frac{2}{5}$ (b) $\frac{3}{5}$ (c) $\frac{4}{5}$
5. (a) $\frac{5}{12}$ (b) $\frac{9}{12}$
6. (b) $\frac{3}{5}$ (c) $\frac{2}{5}$

Exercise 7B

1. 0.48; 0.43; 0.38; 0.41; 0.39
2. (a) 20 (b) 400
4. 12

Exercise 7C

1. (a) $\frac{3}{36}$ (b) $\frac{2}{36}$ (c) 7 (d) $\frac{1}{6}$ (e) $\frac{1}{6}$
2. (a) $\frac{3}{25}$ (b) 0 (c) 4 (d) $\frac{3}{25}$
3. (a) $\frac{1}{16}$ (b) $\frac{1}{16}$
4. (b) $\frac{2}{36}$ (c) $\frac{9}{36}$
5. (b) (i) $\frac{3}{25}$ (ii) $\frac{5}{25}$ (iii) 1
6. (b) $\frac{3}{24}$ (c) $\frac{6}{24}$

Exercise 7D

1. (a) 0.35 (b) 0.65 (c) 0 (d) 0.65
2. (i) $\frac{1}{90}$ (ii) $\frac{1}{10}$ (iii) $\frac{12}{90}$ (iv) $\frac{11}{90}$ (v) $\frac{22}{90}$ (vi) $\frac{1}{90}$ (vii) $\frac{1}{3}$ (viii) $\frac{2}{3}$
3. (i) $\frac{14}{28}$ (ii) $\frac{21}{28}$ (iii) $\frac{8}{28}$ (iv) $\frac{22}{28}$ (v) $\frac{21}{28}$ (vi) 0

Exercise 7E

1. $\frac{1}{12}$; $\frac{7}{12}$
2. (a) 0.04 (b) 0.32 (c) 0.216 (d) 0.784
3. (a) 0.504 (b) 0.006 (c) 0.994
4. (a) 0.216 (b) 0.064 (c) 0.72
5. (a) 0.336 (b) 0.788

Exercise 7F

1. (a) $\frac{9}{25}$ (b) $\frac{4}{25}$ (c) $\frac{13}{25}$ (d) $\frac{12}{25}$
2. (a) $\frac{6}{20}$ (b) $\frac{2}{20}$ (c) $\frac{12}{20}$ (d) $\frac{6}{20}$
3. (c) 0.24
4. (b) $\frac{10}{21}$ (c) $\frac{24}{91}$
5. $\frac{35}{216}$
6. $\frac{4}{15}$

Exercise 7G

1. (a) 0.32 (b) 0.5 (c) 0.8
2. (a) 0.019 (ii) 0.0484 (b) 39.3
3. (a) 0.25 (b) $\frac{1}{3}$ (c) 0 (d) 0

Exercise 7H

1. 0.4
2. 0.2
3. (b) 0.68 (c) 0.35
4. (a) 15 (b) $\frac{29}{40}$

Exercise 7I

1. 32; 0.6
2. (a) 0.8   (b) 500
3. (a) $\frac{1}{3}$ (b) 33.3
4. (a) 278

Exercise 7J

1. 4.5
2. (a) 93   (b) 95
3. (a) 69   (b) 47

Exercise 7K

1. (a) RRR RRB RBR BRR BBR BRB BBB
    (b) (i) $\frac{1}{8}$ (ii) $\frac{3}{8}$ (c) $\frac{1}{8}, \frac{3}{8}, \frac{3}{8}, \frac{1}{8}$
2. (a) (i) $\frac{3}{8}$ (ii) $\frac{1}{2}$
3. (b) $\frac{1}{81}$ (c) (i) $\frac{1}{3}$ (ii) 0.395
4. (a) (i) $\frac{1}{3}$ (ii) $\frac{1}{27}$ (b) $\frac{6}{27}$ (c) $\frac{8}{27}, \frac{12}{27}, \frac{6}{27}, \frac{1}{27}$ (e) positive skew

Revision Exercise 7

1. (a) 0.35, 0.36, 0.37 (b) 0.37
2. (a) 5   (b) B
3. (a) 140   (b) $\frac{9}{28}$ (c) $\frac{3}{7}$ (d) $\frac{1}{15}$
4. (a) $\frac{13}{20}$ (b) $\frac{7}{20}$ (c) 0   (d) 15   (e) 10
5. (a) $\frac{1}{6}$ (b) $\frac{1}{6}$ (c) $\frac{1}{36}$ (d) 6, 6, 6; 5, 5, 6; 5, 6, 5.   (e) $\frac{1}{72}$
6. (a) AB, AC, AD, BC, BD, CD.   (b) $\frac{1}{4}$ (c) $\frac{1}{3}$ (d) $\frac{1}{6}$
7. (a) $\frac{1}{3}$ (b) $\frac{1}{3}$ (c) (i) 3, 4, 5 etc (ii) 10   (d) (i) 3   (ii) $\frac{1}{216}$
8. (a) 35   (b) $\frac{27}{35}$ (d) $\frac{14}{17}$ (e) $\frac{11}{27}$
9. (b) (i) $\frac{1}{6}$ (ii) $\frac{5}{6}$ (iii) 0
10. (a) 0.336   (b) 0.452   (c) 0.212
11. (a) (i) $\frac{3}{4}$ (ii) $\frac{1}{4}$ (b) $\frac{9}{16}$ (c) $\frac{81}{256}$
12. (a) $\frac{6}{25}$ (b) $\frac{23}{50}$ (c) $\frac{13}{23}$
13. (b) $\frac{115}{208}$ (c) 0.1243
14. (a) $\frac{1}{80}$ (c) (i) $\frac{1}{8}$ (ii) $\frac{31}{40}$
15. (a) 8, 9, 10   (b) 1, 1; 1, 2; 2, 1; 2, 2   (c) $\frac{2}{5}$ (e) $\frac{4}{25}$
16. (a) $\frac{2}{5}$ (b) $\frac{3}{5}$
17. (a) $\frac{1}{5}$ (b) 500   (d) (i) 0.1937   (ii) 0.9298
18. (a) 0.3456   (b) 0.2592; 0.0778 (c) 0.0105
19. (a) (i) $\frac{1}{2}$ (ii) 1   (iii) 0   (iv) $\frac{1}{4}$ (b) (i) 22   (ii) liked both   (iii) $\frac{1}{10}$ (iv) $\frac{11}{13}$
20. (b) (i) 0.28   (ii) 0.9895
21. (a) $\frac{6}{7}$ (b) $\frac{25}{49}$ (d) $\frac{1}{21}$
22. (b) (i) 0.15   (ii) 0.27   (c) 162
23. (b) (i) 0.302   (ii) 0.268   (iii) 0.323
24. (a) 0.540   (b) 0.341   (c) 0.0989   (d) 0.0201

Check out 7

1. (a) (i) 0.2   (ii) 0.8   (iii) 0.3   (iv) 0   (b) 0.3
2. (b) (i) $\frac{1}{12}$ (ii) 0   (c) 6
3. 0.26
5. (a) (i) $\frac{7}{36}$ (ii) $\frac{1}{4}$ (iii) 18.5   (b) (i) 0.94208

8 Coursework

Check in 8

1. (i) Colour, shape, ...,   (ii) Bridge length,   (iii) Number of lenses.
2.

	Under 20	20–44	45–64	65 and over
Male				
Female				

Exercise 8A

1. Pocket money is dependent upon age.
2. GCSE grades are a good predictor of A level grades.
3. Book Royalties are dependent upon their length.
4. There is a relationship between age and the colour of cars.
5. There is a relationship between age and ability to estimate length of time.
6. Hand/eye coordination skills can be developed through playing computer games.
7. Life expectancy is greater now than 100 years ago.

Exercise 8B

1. Primary, conduct an experiment.
2. Primary, carry out a survey; Secondary, look at records of births.
3. Primary, conduct a taste test experiment.
4. Secondary, records of fatal road traffic accidents.
5. Secondary, GNP, population, land mass, number of doctors, . . .
6. Secondary, past meteorological records.

Exercise 8C

1. Two-way tables (for different eras) with length of track & type of music.
2. Table with headings height, arm span, hand span, wrist circumference.
3. Table with headings age, number recalled.
4. Two-way table with days of the week & number per table. (Collect number of tables booked).
5. Table with headings brand X, brand Y, ...
6. Two-way tables (for blonde and non-blondes) with age & level of higher education / career.

Check Out

1. Wrist circumference is related to thumb circumference.
2. Primary data, wrist and thumb circumference, survey.
3. Scatter graphs, box & whisker back to back stem & leaf, mean & standard deviation, median & inter-quartile range.
4. Sort and compare data by gender and by age group, for example.

9 ICT in Statistics

There are no answers given to the exercises. You can check the answers to most of the questions set in the answers given above.

Exam practice paper: Higher tier

1. (a) Line scaled, 6 points plotted; (b) Steady increase, check machinery or operator
2. (a) $\frac{72}{100}, \frac{75}{100}, \frac{55}{100}$; (b) two box & whisker diagrams; (c) girls quicker & middle 50% more spread out.
3. (a) Find typical response/check wording; (b) Will be leaving/do not wear uniform; (c) 22, 22, 19, 19, 18; (d) 33.5, 33.5, 28, 28, 27 first two groups need rounding up, but that would give one too many and we have no reason to round one up and the other down.
4. 6.18 cm.
5. (a) $\frac{4}{5}$ (b) Tree diagram   (c) $\frac{19}{35}$.
6. (a) 0.142; (b) No correlation; (c) Change, no, only marginal difference.
7. (a) Plot (2, 0.81) (4, 1.69) (6, 2.56) (8, 3.24) (10, 4) (12, 4.84) (15, 6.25) (20, 8.41); (b) Line through (0, 0); (c) m = 0.85.
8. (a) A D B B C D D B C B A B A D D C C C B C; (b) $\frac{5}{20}$; (c) $\frac{13}{20}$; (d) Very different; need to take larger sample size.
9. (a) Frequency density 0.6, 1.2, 3.6, 1.15, 0.8, 0.1; (b) (i) 189.2; (ii) 18.45; (c) 112 kmph, less than 10% of cars travelling faster than this.
10. (a) 1.367, 0.856; (b) First class ~ 1 day quicker, spread about the same.
11. (a) 3; (b) CF 16, 51, 76, 94, 108, 117, 120; (c) 2.
12. (a) 107.9; (b) 102; (c) Increase by 2%.
13. (a) 0.25; (b) Fixed number of trials, only two outcomes; (c) 0.0879; 0.0146; 0.000977; (d) 5.
14. (a) Plot data; (b) 87.6, 88.325, 89.4, 90.625, 91.975, 93.175, 94.4; (c) trend line; (d) ~98.1; (e) ~ −3.7; (f) ~ 95.6.

Exam practice paper: Foundation tier

1. (a) Length; (b) Number of spots; (c) Colouring.
2. X very close to zero; Y ~ 0.7; Z ~ 0.1.
3. (a) £5600; (b) £8000; (c) Angles 252°, 36°, 45°, 27°.
4. (a) 3; (b) M; (c) 7.
5. (a) Already buying fish and chips, all one place; (b) Random sample, put names in hat and pick required number; (c) Leading; (d) Non leading question with tick box choices; (e) Check wording, find typical responses.
6. (a) 107.9; (b) Housing; (c) £2424.40; (d) £320.
7. (a) 65–80; (b) CF 12, 30, 108, 162, 185, 197, 200; (c) (i); (ii); (d) 112 kmph as less than 10% travelling faster than this speed.
8. (a)

	5	6	7	8	9
Across	5	4	3	1	0
Down	5	4	2	0	1

(b) 4; (c) 6;

(d) Cryptic has longer words on average and more varied length of answer.
9. Do not use thick lines; Begin scale at zero.
10. (a) High response rate, Expensive; (b) Use 001–500 for pages choose required amount. Give names numbers from 501–850 choose amount required, ignore repeats and greater numbers.
11. (a)

```
| 8 | 2 3              Key: 3|1 means 3.1 seconds
| 7 | 2 2 2 3 5 7
| 6 | 0 1 9
| 5 | 2 5
| 4 | 4
| 3 | 1
```

(b) 7.2 7.5, 5.5;
(c) B & W leaves 3.1–8.3;
(d) Negative;
(e) Girls were quicker, but times more varied.
12. (b) Negative; (c) January 2000; (d) February – April 2001; (f) Time would eventually become zero.
13. (a) $\frac{1}{4}$; (b) 50; (c) (i) $\frac{2}{5}$ (ii) $\frac{3}{10}$ (iii) $\frac{5}{20}$; (d) No, same probability as guess.
14. (a) Data logging; (b) Plot (5, 130) (10, 134) (15, 142) (20, 144) (25, 146) (30, 148) (35, 145) (40, 140); (c) 10–15 minutes; (d) More readings, may have gone above 150 between 25–30 or 30–35 minutes.
15. (c) ~ £90 – 91 million; (d) Christmas purchases.

Index